T0281275

A Gardner's Workout

A Gardner's Workout
Training the Mind and Entertaining the Spirit

Martin Gardner

CRC Press
Taylor & Francis Group
Boca Raton London New York

CRC Press is an imprint of the
Taylor & Francis Group, an **informa** business

AN A K PETERS BOOK

First published 2001 by A K Peters, Ltd.

Published 2019 by CRC Press
Taylor & Francis Group
6000 Broken Sound Parkway NW, Suite 300
Boca Raton, FL 33487-2742

© 2001 by Taylor & Francis Group, LLC
CRC Press is an imprint of Taylor & Francis Group, an Informa business

First issued in paperback 2019

No claim to original U.S. Government works

ISBN 13: 978-0-367-45516-3 (pbk)
ISBN 13: 978-1-56881-120-8 (hbk)

Visit the Taylor & Francis Web site at
http://www.taylorandfrancis.com

and the CRC Press Web site at
http://www.crcpress.com

Library of Congress Cataloging-in-Publication Data

Gardner, Martin
    A gardner's workout : training the mind and entertaining the spirit /
Martin Gardner.
        p. cm.
    ISBN 1-56881-120-9
    1. Mathematical recreations. I. Title.

QA95 .G158 2001
793.7´4--dc21   2001021427

To all the underpaid teachers of mathematics,
everywhere, who love their subject
and are able to communicate that love
to their students.

Table of Contents

Preface

For 25 years I had the honor and pleasure of writing the "Mathematical Games" column in *Scientific American*. All those columns have now been reprinted, with updating, in fifteen volumes, starting with *The Scientific American Book of Mathematical Puzzles and Diversions* and ending with *Last Recreations*.

Since I stopped writing the column I have from time to time contributed articles and book reviews about mathematics to both academic journals and popular magazines. Forty-one of these pieces are gathered here. The most controversial is the final review in which I criticize a current teaching fad known as the "new new Math."

By the time this book is published I would guess and hope that new new math is being abandoned almost as rapidly as the old new math faded. I could be wrong. In any case, it may be decades before our public education is able to attract competent teachers who have learned how to teach math to pre-college students without putting them to sleep. There are, of course, many teachers who deserve nothing but praise. It is to them I have dedicated this book.

Martin Gardner
Hendersonville, NC

Part 1

Chapter 1
The Opaque Cube

The Opaque Cube

I want to propose the following unsolved problem. As far as I know, I am the first to ask it.

⋆ ⋆ ⋆

What is the minimal area of surfaces inside a transparent cube that will render it opaque?

⋆ ⋆ ⋆

By opaque I mean *if the surfaces are opaque, no ray of light, entering the cube from any direction, will pass through it.*

The answer may or may not be the *minimal surface* spanning the twelve edges of the cube. This question also is unanswered. See the discussion of it by Courant and Robbins ([1], Ch. 7).

That the minimal spanning surface may not be the answer to the opaque cube problem is suggested by the fact that the minimal Steiner tree spanning the four corners of a square is not the answer to the opaque square problem. The best known solution for the square (also not proved minimal) is shown in Figure 1. The square problem is discussed by Ross Honsberger ([2], p. 22).

My best solution for the cube is to join the center to all the corners. These lines outline 12 triangles with a combined area of $3\sqrt{2}$.

I believe the opaque cube problem to be extremely difficult. It is keeping me awake at night![1]

---

[1]The note appeared in the Dutch periodical *Cubism for Fun* (No. 23, March 1990, p. 15). I followed this with a second note, "The Opaque Cube Again," in the same periodical (No. 25, December 1990, Part 1, p. 14).

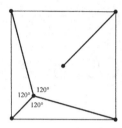

Figure 1. Total length = 2.639+.

References

[1] R. Courant and H. Robbins. *What is Mathematics?*, 4th edition, Oxford University Press, 1947.

[2] R. Honsburger. "Mathematical Morsels," The Mathematical Association of America, 1978.

The Opaque Cube Again

So far as I know, no one has yet proved that the solution for the opaque square is minimal, nor has anyone proved a minimal surface spanning the edges of a cube.

I refer to the problem posed in CFF 23 (March 1990), p. 15: "Find the least-area surface that will block any light ray trying to pass through a unit cube."

From Stephen Harvey, Dunedin, New Zealand, I received a calculation of the area of a surface such as pictured in Figure 240 of Courant and Robbins "*What Is Mathematics?*". This area ($A = 4.2425$) is slightly less than $3\sqrt{2}$.

From H.S.M. Coxeter, Toronto, Canada, I received a letter in which he shows that the spanning surface of Courant and Robbins cannot have straight edges, which makes this area very difficult to compute.

Even this surface can be improved slightly, however, as is shown by Kenneth Brakke, Mathematics Department, Susquehanna University, Selingsgrove, Pennsylvania. He has a marvelous computer program that searches for minimal surfaces. He found a minimal surface spanning the cube's edges that has an area of $A = 4.2324$. This is the best solution yet for the Opaque Cube.

Opaque Cubes by Kenneth Brakke

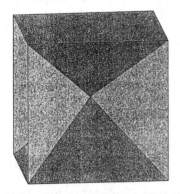

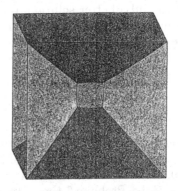

Figure 2. Twelve triangles from the edges to the centre. Area $= 3\sqrt{2} \approx$ 4.2426.

Figure 3. Soap film formed by dipping a cubical frame in a soap solution. Area $\approx$ 4.2398.

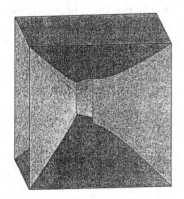

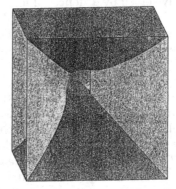

Figure 4. Generalization of opaque square solution to three dimensions. Topologically, this is the opaque square solution stretched vertically with the top and bottom faces of the cube added. Area $\approx$ 4.2343.

Figure 5. My best solution. This is like Figure 4, except it has three-fold symmetry in place of two-fold. Area $\approx$ 4.2324.

Kenneth Brakke will give a fuller treatment of his results elsewhere. Meanwhile he kindly allowed the editors of CFF to publish some of his computer printouts (see Figures 2–5). If they inspire any reader to find a better solution, he would like to hear about it.

At the moment it seems unlikely that the minimum solution for the opaque cube will consist of (disconnected) surfaces that do not span the cube's edges with a single surface. However, this is far from proved.

Postscript

The picture of Brakke's best solution to the opaque cube, reproduced here as Figure 5, made the cover of *The American Mathematical Monthly* Vol. 99, (November 1992). The cover illustrated Brakke's paper, "The Opaque Cube Problem," that ran in Richard Guy's "Unsolved Problems" column. Brakke also discusses the opaque sphere problem.

The opaque square, with its single Steiner point, obviously generalizes to opaque regular polygons. The limiting case, as the number of polygon sides increases, is, of course, the circle. The opaque circle is probably solved by a "fence" of length 2π that is the limit of an infinite series. The solver is Bernd Kawohl, a professor of mathematics at the University of Cologne's Mathematical Institute. His solution is given in "The Opaque Square and the Opaque Circle," in the proceedings of a conference in Oberwolfach, reprinted in the *International Series of Numerical Mathematics*, Vol. 123 (1997), pp. 339–346.

Figure 6 and 7 show the best known solutions for the opaque pentagon and hexagon. Note that each is totally lacking in symmetry. Kawohl's conjectured solution for the opaque circle is based on extrapolating from the polygonal cases. See also Kawohl's article "Symmetry or Not?", in *The Mathematical Intelligencer*, Vol. 20 (1998), pp. 16–22, in which both the square and circle cases are discussed.

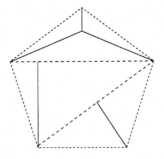

Figure 6. The opaque pentagon. Length ≈ 3.528.

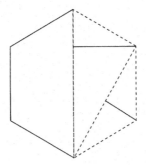

Figure 7. The opaque hexagon. Length = $(7 + \sqrt{3})/2 \approx 4.366$.

Chapter 2
The Square Root of 2 = 1.414 213 562 373 095 ...

Roses are red,
Violets are blue.
One point 414 ...
Is the square root of two.

I confess that I wrote the above jingle only to have some light verse top this article. The dots at the end of the third line indicate that the decimal fraction is endless and nonrepeating. In other words, $\sqrt{2}$ is irrational. Although its decimal digits, like those of other famous irrationals such as π and e, *look* like a sequence of random digits, they are far from random because if you know what the number is you can always calculate the next digit after any break in the sequence. Such irrationals also should not be called "patternless" because they have a pattern provided by any formula that calculates them. The square root of two, for example, is the limit of the following continued (endless) fraction:

$$\sqrt{2} = \cfrac{1}{2 + \cfrac{1}{2 + \cfrac{1}{2 + 1}}}$$

From this continued fraction one can derive rational fractions (fractions with integers above and below the line) that give $\sqrt{2}$ to any desired accuracy. The sequence 1/1, 3/2, 7/5, 17/12, 41/29, 99/70, 239/169, 577/408, 1303/985 ... is sometimes called "Eudoxus' ladder" after an ancient Greek astronomer and geometrician. The fractions are alternately higher and lower than their limit, which is $\sqrt{2}$. Each fraction is closer to $\sqrt{2}$ than its predecessor. The best approximation with numerator and denominator not exceeding three digits is 577/408. It gives

This article first appeared in *Math Horizons* (April 1997).

9

$\sqrt{2}$ to five decimal places. If a fraction in this sequence is represented by a/b, the next fraction will be $(a + 2b)/(a + b)$. Note than on each "rung" of the ladder the numerator is the sum of its denominator and the denominator of the preceding fraction.

David Wells, in his *Penguin Dictionary of Curious and Interesting Numbers* (pages 34–35) gives some strange properties of the multiples of $\sqrt{2}$. For example, write in a line the multiples, omitting the fractional part. For example, 1 times $\sqrt{2}$, ignoring the decimal digits is 1. Twice $\sqrt{2}$, ignoring the decimals, is 2. In this way you obtain the following sequence: 1, 2, 4, 5, 7, 8, ...

Beneath this sequence put down the numbers *missing* from the first sequence:

1	2	4	5	7	8	9	11	12	...
3	6	10	13	17	20	23	27	30	...

The difference between the top and bottom numbers at each nth position is always twice n.

Normal Numbers

Any n^{th} root of a positive integer (in all that follows "integer" will mean a positive integer) not itself an nth power is irrational. Although all such irrational roots have decimal digits that are neither random nor patternless, they are all, so far as anyone knows, "normal." This means that if you specify any pattern of digits, such as a single digit, pairs of adjacent digits, triplets of adjacent digits, and so on, in the long run the pattern will appear with just the frequency you would expect on the assumption that the probability of finding any given digit at any given place is always 1/10.

The pattern need not involve adjacent digits. They can be spaced any way you like. For example, you might consider the pattern abc, where a and b are separated by, say, seven digits, and b and c are separated by, say, 100 digits. All tests so far to determine the frequency of such patterns have shown that all irrational roots, in any base notation, are normal.

The most extensive tests for the normalcy of certain irrationals have been made for π because π has now been calculated to hundreds of millions of digits, but similar tests of other famous irrationals such as e and the golden ratio have shown no deviations from normalcy. I do not know how far $\sqrt{2}$ has been calculated, though I have a reference

(*TIME*, October 25, 1971) to it having been carried to more than a million digits in 1971 by Jacques Dutka, then a Columbia University mathematician.

One might imagine that all irrationals are normal, but it is easy to see that this is not the case. A popular example is the binary fraction .10100100010000 .... The number clearly is not rational and just as clearly is far from normal.

√2 and Drowning at Sea

The discovery of irrational roots was first made by the Pythagoreans, a secret brotherhood that flourished in ancient Greece. Their discovery of the first irrational number, the square root of 2, was a milestone in the history of mathematics. In geometrical form this says that the diagonal of a unit square is incommensurable with the square's side. No ruler, no matter how finely graduated, can accurately measure the two line segments. If the side of a square is rational, the diagonal will be irrational, and vice versa.

There are two legends about the explosive effect of this discovery. One is that a Pythagorean named Hippasus was sworn not to reveal the discovery because it shattered the Pythagorean belief that integers accurately measure all things. Hippasus broke the vow. As a result he was drowned at sea either by suicide, murder, or the wrath of the gods—the legend has many variations. The other legend has the Pythagoreans celebrating their great discovery by sacrificing many oxen to the gods. The discovery of incommensurable line segments had a profound influence on Plato, who wrote about it in his *Laws*.

Infinite Descent

The Greeks proved the incommensurability of a square's side and diagonal by a clever "infinite descent" proof using the diagram shown in Figure 1.

Assume that the side of the largest square is commensurable with its diagonal. If so, each of the two line segments will be multiples of a unit that we call k. Draw a smaller square of side b, choosing point x so that $a = c$. Side b of this square will be commensurable with its

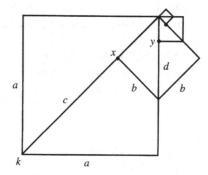

Figure 1. An infinite descent proof that $\sqrt{2}$ is irrational.

diagonal because each is a multiple of k. Next we select point y so that $d = b$. Again, the side and diagonal of this smaller square will be commensurable with respect to k.

This process can be continued to infinity as suggested by the fourth tiny square. The sides of all these squares cannot be zero, but at some point in the endless construction we reach a square with a side less than k. A length less than k cannot be a multiple of k, so we have encountered a contradiction proving that our assumption, that the side and diagonal of a square are commensurable, is false. If the square's side is 1, the diagonal is $\sqrt{2}$. We have shown that $\sqrt{2}$ is irrational.

We can express the proof another way. We seem to get an infinite series of integers (multiples of k), each smaller than the previous one, but such a series obviously must be finite.

Hugo Steinhaus, in the first chapter of *Mathematical Snapshots*, gives a different geometrical proof by infinite descent. It is based on the rectangle shown in Figure 2. Its sides are in a ratio such that if the rectangle is sliced in half as shown, each half will be a rectangle similar to the original one. If the sides are labeled as indicated, a and b will be in the same ratio as $a/2$ and b. The equation reduces to $a^2 = 2b^2$, so if $b = 1$, a will be $\sqrt{2}$.

Assume that a and b are commensurable, each side a multiple of unit k. Of course, k can be any unit, inches, centimeters, or whatever.

In Figure 3 we have attached to the long side of rectangle ab a congruent rectangle that has been given a quarter turn clockwise. This produces a larger rectangle of sides b and $(a + b)$. By cutting two squares of side b from this large rectangle we produce the smaller

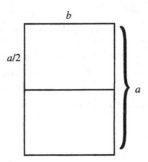

Figure 2. Another infinite descent proof.

shaded rectangle. Its sides are b and $(a-b)$. Because a and b are integers, $(a-b)$ must also be an integer. Therefore the shaded rectangle must have sides that are multiples of k.

We can repeat the procedure by cutting two squares from the shaded rectangle to create a still smaller rectangle, similar to the shaded one, with sides that also must be multiples of k. As in the previous proof, if this process is continued we soon produce a rectangle with sides smaller than k. We have reached a contradiction. The procedure can be carried to infinity, but one cannot have an infinite sequence of integers that keep getting smaller and smaller. Therefore a and b are incommensurable, and $\sqrt{2}$ is irrational. Infinite descent proofs can be given algebraic forms, many of which generalize to proving that any n^{th} root not an nth power is irrational.

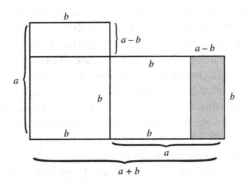

Figure 3. Steinhaus' infinite descent proof.

For an application of the $1 \times \sqrt{2}$ rectangle to a magic trick involving the repeated folding of a playing card, see the chapter on rep-tiles in my *Unexpected Hanging and Other Mathematical Diversions*. The rectangle is called an order-2 rep-tile because it can be cut into two parts each similar to itself. British and European sheets of paper usually have sides in a 1 to $\sqrt{2}$ ratio so that when halved, quartered, and so on, the sheets remain similar.

Odds and Evens

The ancient Greeks also had an elegant way of using the laws of odd and even numbers to prove $\sqrt{2}$ is irrational. It can be expressed in numerous ways, but the following seems the simplest.

Let a stand for the hypotenuse of a right isosceles triangle and b for its side. We know from the Pythagorean theorem that $a^2 = 2b^2$, or $a^2/b^2 = 2$. The fraction a/b obviously is between 1 and 2. Assume it is reduced to lowest terms—that is, its top and bottom numbers have no common divisor other than 1. We know b is greater that 1, otherwise a/b would be an integer.

The right side of $a^2 = 2b^2$ is even, therefore the left side a^2 is also even, and a is even because the square root of any even number is even. For a we can substitute $2x$ where x is any integer. Squaring $2x$ gives $4x^2$, so we can write $4x^2 = 2b^2$. This reduces to $2x^2 = b^2$. The left side is even, therefore b^2 is even and b is even. Because both a and b are even, each can be divided by 2. This contradicts the assumption that a/b has been reduced to lowest terms. We have proved that a/b cannot be a rational fraction between 1 and 2, therefore $\sqrt{2}$ is irrational.

Euclid gave this proof in Book 10, and Aristotle alludes to it in many places. According to Plato in his dialogue *Theaetetus* (section 147), Theodorus of Cyrene, a brilliant philosopher and geometrician, also proved the irrationality of the square roots of all nonsquares of 3 through 17. Alas, none of his writings survive, so we don't know how he did it, or why he stopped at 17. Incidentally, Theodorus was banished from Cyrene because he doubted the existence of the Greek gods. With suitable modifications, parity (odd–even) proofs of $\sqrt{2}$ can be generalized to all n^{th} roots of integers that are not n^{th} powers.

Each of the foregoing proofs is a *reductio ad absurdum* or "indirect" proof in which an assumption is made then later proved false by a contradiction. A whimsical indirect proof of the irrationality of $\sqrt{2}$ is based on the final digit of square numbers. It is easy to see that this digit must be 0, 1, 4, 5, 6, or 9. Consider again the equation $a^2 = 2b^2$, where a/b is reduced to lowest terms, b greater than 1.

The terminal digit of both a^2 and b^2 must be one of the six listed above. On the right side of $a^2 = 2b^2$, b^2 is multiplied by 2, therefore the final digit of $2b^2$ must be 0, 2, or 8. It cannot be 2 or 8 because there is no 2 or 8 as the last digit of a^2. The only match is 0. So a^2 and $2b^2$ must each end in zero. It follows that a must end in 0, and b must end in 0 or 5. In either case both a and b are divisible by 5, contradicting the assumption that a/b is reduced to lowest terms. Hence a/b is irrational and $\sqrt{2}$ is irrational.

Similar terminal digit proofs of the irrationality of $\sqrt{2}$ can be formulated in other base notations. In binary notation, for example, the proof is unusually simple. The left side of $a^2 = 2b^2$ terminates in an even number of zeros and the right side terminates in an odd number of zeros.

Many elegant proofs of the irrationality of $\sqrt{2}$ are based on the fundamental theorem of arithmetic, which states that every integer is the product of a unique set of primes. Here is one of the easiest to follow.

As before, we use the equation $a^2 = 2b^2$ where a/b is a rational fraction reduced to lowest terms, b greater than 1. The term a^2 must have an even number of prime factors. Why? Because if a is the product of either an odd or an even number of primes, its square will have twice as many prime factors.

Consider now the right side of $a^2 = 2b^2$. It will have an odd number of prime factors because to the even number of prime factors of b^2 we add the prime factor 2. We have produced a contradiction because the number of prime factors for the two sides of the equation cannot be even on one side and odd on the other. It is not difficult to see that the proof applies to the square root of any prime, or to an integer with an odd number of prime factors.

Prime divisors provide a simple proof that any square root not an integer is irrational. We apply it first to $\sqrt{2}$. From $a^2 = 2b^2$ we can derive the equation $b^2 = a^2/2$ which is the same as a times $a/2$. If a prime divides the product of two integers x and y, it obviously must divide either x or y. Let a^2 and a be the two integers whose product is $a^2/2$. There must be a prime that divides b^2 because b is greater than 1. This same prime must divide the right side of the equation, therefore it must divide $a/2$ or a. In either case it divides a because if it divides half of a, it will also divide a. Contradiction! We have shown that a prime divides both a and b, therefore a/b cannot be a rational fraction reduced to lowest terms.

Substitute for 2 any integer whose square root is not an integer and the foregoing proof holds. With further generalizations the proof will apply to all n^{th} roots of integers that are not n^{th} powers.

Another simple proof of the irrationality of $\sqrt{2}$ is based on inequalities. If a/b is $\sqrt{2}$ reduced to lowest terms, then b is less than a, and a is less than $2b$, therefore $(a - b)$ is less than b. Start with $a^2 = 2b^2$, and make the following changes:

$$a^2 - ab = 2b^2 - ab$$
$$a(a - b) = b(2b - a)$$
$$a/b = (2b - a)/(a - b)$$

As we have seen, $(a - b)$ is smaller than b. We have contradicted the assumption that a/b is reduced to lowest terms. This proof also generalizes to any n^{th} root of any number not an nth power.

There are dozens of other ways to prove the irrationality of the square roots of integers that are not squares, many of which extend easily to n^{th} roots. They all come down to the following theorem: If a/b is a rational fraction in lowest terms, b greater than 1, then any power of a/b will also be a rational fraction that cannot be reduced to lower terms.

This can be proved by the following argument involving prime factors. Assume that a/b, with b greater than 1, is reduced to lowest terms. The prime factors of a will have no factors in common with b, otherwise the common factors cancel out and a/b is reduced. Consider now the square of a/b. The factors above the line will be the same as before, each repeated twice, and the same for the prime factors below the line. There are still no common factors to cancel. This means that the square of a rational fraction reduced to lowest terms is another fraction reduced to lowest terms, so it cannot be an integer. In brief, no integer not a square can have a square root that is rational.

The argument obviously applies to cubes and all higher roots. For example, a^3/b^3 is $(a \times a \times a)/(b \times b \times b)$. This too is a nonreducible fraction because there are no common prime factors above and below the line to be canceled. Is there any simpler, easier to comprehend, way to show that n^{th} roots of integers not n^{th} powers are irrational?

When I was in high school and first learned that $\sqrt{2}$ could not be expressed as a rational fraction, I couldn't believe it. I squandered many hours in study periods trying to find such a fraction. Eventually I convinced myself it couldn't be done, but today I have no memory of how I proved it, if indeed I did. I like to think it was one of the proofs given in this article. It would be interesting to know how many mathematicians, far greater than I, had a similar experience when they were very young.

Note that all the proofs in this article are *reductio ad absurdem* proofs. They illustrate how powerful this type of proof is. As G. H. Hardy put it in his famous *Mathematician's Apology*:

> It is a far finer gambit than any chess gambit: a chess player may offer the sacrifice of a pawn or even a piece, but a mathematician offers *the game*.

References

Hundreds of books contain proofs of the irrationality of $\sqrt{2}$ and more general proofs of the irrationality of any n^{th} root of an integer not an nth power. What follows are references in easily accessible periodicals.

[1] Beckenbach, Edwin. "On the Positive Square Root of Two." *Math. Teacher* 62, April 1969, 261–267.

[2] Bloom, David. "A One-Sentence Proof that $\sqrt{2}$ is Irrational." *Math. Mag.* 68, Oct. 1995, 286.

[3] Bumcrot, Robert. "Irrationality Made Easy." *College Math. J.* 17, May 1986, 243–233.

[4] Estermann, T. "The Irrationality of $\sqrt{2}$." *Math. Gazette* 59, June 1975.

[5] Fine, Nathan. "Look Ma, No Primes." *Math. Mag.* 49, Nov. 1976, 249. See also letters in April 1977, 175.

[6] Goodstein, R. L. "The Irrationality of a Root of a Non-Square Integer." *Math. Gazette* 53, Feb. 1969, 50.

[7] Harris, V. C. "Terminal Digit Proof that $\sqrt{2}$ Is Irrational." *Math. Gazette* 53, Feb. 1969, 65.

[8] Harris, V.C. "On Proofs of the Irrationality of $\sqrt{2}$." *Math Teacher* 64, January 1971, 19–21. See also his "$\sqrt{2}$ Sequel," 64, Dec. 1971, 760.

[9] Lange, L. J. "A Simple Irrationality Proof for nth Roots of Positive Integers." *Math. Mag.* 42, November 1969, 242–243.

[10] Lindstrom, Peter. "Another Look at $\sqrt{2}$." *Math. Teacher* 72, May 1979, 346–347.

[11] Maier, E.A., and Ivan Niven. "A Method Of Establishing Certain Irrationalities." *Math. Mag.* 37, Sept./Oct. 1964, 208–210.

[12] Randall, T.J. "$\sqrt{2}$ Revisited." *Math. Gazette* 67, December 1983, 442.

[13] Rothbart, Andrea. "Back to $\sqrt{2}$." *Math. Teacher* 65, November 1972, 667–668.

[14] Shibata, Toshio. "On a Proof of the Irrationality of $\sqrt{2}$." *Math. Teacher* 67, Feb. 1974, 119.

[15] Strickland, Warren. "A More General Proof for $\sqrt{2}$." *Math. Teacher* 65, February 1972, 109.

[16] Subbarao, M. V. "A Simple Irrationality Proof for Quadratic Surds." *Amer. Math. Monthly* 75, Aug./Sept. 1968, 772–773.

[17] Waterhouse, William. "Why Square Roots Are Irrational." *Amer. Math. Monthly* 93, March 1986, 213–214.

[18] Zoll, Edward. "A Fourteenth Proof for $\sqrt{2}$." *Math. Teacher* 65, Jan. 1972, 30.

Postscript

Here are two short indirect proofs, not in my article, that the square root of 2 is irrational.

1. If $\sqrt{2}$ is a rational fraction there must be a smallest positive integer k that would make $k\sqrt{2}$ an integer. But $k\sqrt{2} - k$ is a *smaller* such integer. Contradiction.

2. As made clear in this chapter, if the square root of 2 is an integral fraction between 1 and 2, the equation $a^2 = 2b^2$ would have a solution in integers. In base-3 notation, the last non-zero digit of a square must be 1. This applies to the left side of the equation. But the last non-zero digit of $2b^2$ is 2, a contradiction proving that $a^2 = 2b^2$ has no solution in integers.

Monte Zerger called my attention to an amazing way to generate all the convergents of the square root of 2. He found this explained in a paper by D. V. Anderson in *The Mathematical Gazette* (November 1996, pp. 574–575). Start with the sequence of powers of 2, each power appearing twice, then add adjacent digits in the manner of Pascal's triangle. The first two entries of each row are the successive convergents that I listed earlier as the rungs of Eudoxus' ladder!

```
1   1   2     2      4      4     8     8
    2   3    4      6      8     12    16
        5   7    10     14     20    28
           12   17    24     34    48
              29   41    58    82
                 70   99   140
                    169   239
```

I was surprised and honored when the foregoing article won the 1998 Trevor Evans Award of $250, given each year to articles published in *Math Horizons*. Trevor Evans was a distinguished mathematician at Emory University.

Chapter 3
Flip, the Psychic Robot

Can You Outwit a Mindless Automaton?

Here's a chance to take on an opponent who has no control over the moves he makes. Still, winning may be tougher than you think.

The game is matching pennies. You flip a coin and then Flip, the robot, will do the same. If Flip's flip matches yours (heads after you have thrown heads, or tails after your tails), Flip wins. If not (tails after your heads, or heads after your tails), you win. Flip has challenged you to a 25-game match.

Here's how it works. For each game (1–25), throw a coin and note whether you threw heads or tails. Being a mere sheet of paper, Flip can't toss for himself, so you now get to tell him what to throw. Thus, if you threw heads, you may want to tell Flip to throw tails. But there's a catch: Sometimes Flip will obey your commands, sometimes he won't. You have no way of knowing in advance when he'll obey and when he'll disobey.

Once you've chosen heads or tails for Flip (and not before), follow a line that leads from that response all the way to its end to discover Flip's true flip. In following the lines, you may not change directions at intersections. Check the result of each game before proceeding to the next.

Keep a running total of the number of games won by you and the number won by Flip.

You may be surprised to find that every time you play this game Flip will win. Here's why:

The game is designed to counter the way people make choices when trying to beat a machine. Research has shown that most people in such a situation fall into a predictable psychological pattern; the game uses this pattern in designing the robot's responses (whether he'll obey or

This article first appeared in *Games* Magazine (October 1984).

disobey) in a way that works to Flip's advantage. In effect, Flip is psyching out humans trying to psych out Flip.

The game is based on actual experiments in artificial intelligence. In 1969, Soviet researchers programmed a computer to play a game equivalent to this coin-matching test and found that nearly three-quarters of the 61 humans they tested lost. If a human player did not try to psych out the robot but simply chose Flip's responses at random, he or she would have a 50–50 chance of winning.

I designed Flip after reading about the Russian computer program in Russ Walter's *The Secret Guide to Computers* (Birkhauser Boston, 1983).

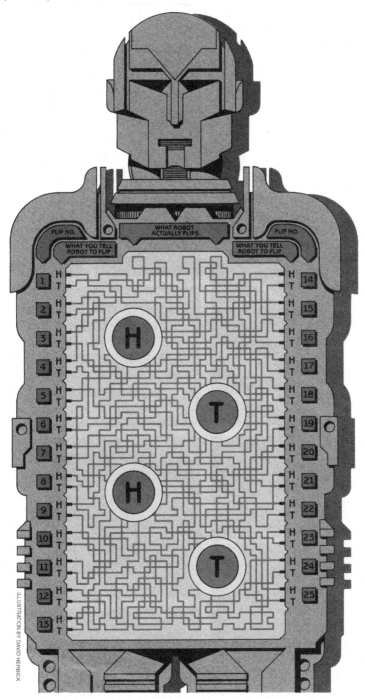

Figure 1. Flip, the psychic robot.

Chapter 4
The Propositional Calculus with Directed Graphs

Many formal logics can be represented by geometrical diagrams that are useful in two ways: They are visual aids that help students grasp the structure of statements in the logic, and in some cases the diagrams can be manipulated in such a way that theorems can be proved and problems solved as efficiently as with algebraic techniques. (For a history of logic diagrams see Gardner [1].)

The first good way to diagram the propositional calculus was by using Venn circles, proposed in 1880 by the British logician John Venn. He discussed his system at much greater length in *Symbolic Logic* [2], where he applied it almost exclusively to a class interpretation of the new Boolean algebra, especially to the diagramming of traditional syllogisms. Because Venn circles apply just as accurately to the binary relations of the propositional interpretation of Boolean algebra, it is curious that even today introductory textbooks of formal logic limit Venn diagrams exclusively to class-inclusion logic.

Figure 1 shows how a pair of Venn circles are shaded to represent each of the connectives of the propositional calculus. (Traditional symbols for the connectives are also shown.) Venn allowed the entire area outside his circles to represent the class consisting of the negations of all terms, but is best to confine this class to a small circle as indicated. This makes the region much easier to shade. Any diagrammed statement can be changed to its negation by exchanging the white and dark areas; the negated diagram is simply the photographic "negative". Three circles will handle all binary relations that concern three or fewer terms, each circle standing for a proposition that must be either true or false. The system is readily extended to diagram for

This article, coauthored with the noted graph theorist Frank Harary, first appeared in *Eureka* (March 1988).

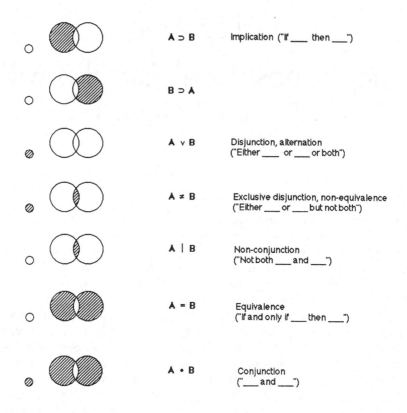

Figure 1. Venn diagrams for binary connectives in the propositional calculus.

four or more terms by using closed curves other than circles, and to matrix forms proposed by John Marquand, Lewis Carroll, and others.

A method of diagramming the propositional calculus with undirected networks was proposed by Gardner [1] in the late 1950s. In the early 1960s Dr. Garrit M. Mes, a Dutch-born surgeon at the Medical Center, Krugersdorp, Africa, improved the network method by adding arrows to its lines. His system was never published, though it is briefly mentioned by Gardner [1]. In 1977 Frank Harary independently thought of the same technique. The purpose of this paper is to explain the method. The authors believe it is a useful educational device. Not only does it diagram statements in the propositional calculus in a way that permits the efficient solving of elementary problems, but

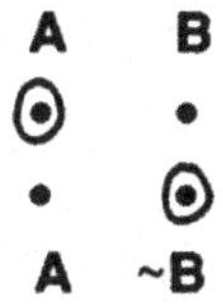

A B

⊙ •

• ⊙

A ~B

Figure 2. **A** is true, **B** false.

it solves them in a way that Charles Peirce liked to call "iconic"—that is, in a manner that strongly resembles the formal structure being analyzed. Moreover, it makes use of diagrams closely related to the networks of logic circuits in today's computer chips.

The digraph system has another great advantage over Venn diagrams. After diagramming the premises of a problem in the propositional calculus, using Venn circles, it is impossible to distinguish the shading of one premise from the shading of another. This makes it extremely difficult to experiment with the problem structure by altering its premises to see what the change entails. In the digraph system, as we shall see, each premise has a diagram isolated from the others. This makes for great ease in exploring the total structure of the problem, seeing how it changes when any premise is removed or altered, or new premises are added.

The fundamental diagram is simple. The two possible values of any term **X** are indicated by two points, one above the other. By convention, the top point is **X** and the bottom point is its negation, ~ **X**. The truth value of a term, when it is known, is indicated by drawing a tiny circle around the appropriate point, as shown in Figure 2. The upper left circle indicates that the proposition labeled **A** is true. The lower right circle shows that proposition B is false—that is, B's negation is affirmed.

The binary connectives are indicated by directed lines as shown in the first column of Figure 3. The second column shows the same diagrams, but simplified by replacing each double line (arrows going both ways) with a single undirected line. It is assumed that an undirected line can be traversed in both directions. If the truth value of any term is known, we can travel from the circled term along any line attached to it, provided the line is undirected or an arrow permits it, to another term that can then be circled.

Two examples will make this clear. Consider the conjunction "**A** and **B**". It asserts that statements **A** and **B** are both true. As the

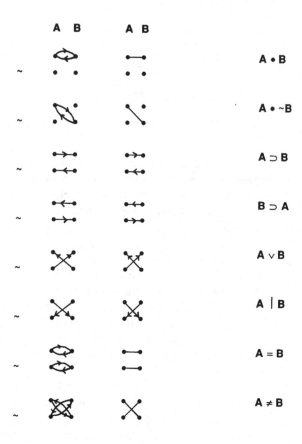

Figure 3.

diagram for this connective indicates, if either **A** or **B** is circled, we can "travel" along the line to the other point and circle it. The diagram for implication, "If **A**, then **B**", shows at once that if **A** is circled we may follow the directed line to **B** and circle it also, but we cannot go against the arrow from **B** to **A**. Similarly, if ~ **B** is circled we may go to ~ **A** and circle it, but we cannot go from ~ **A** to ~ **B**. In other words, knowing **B** is true tells us nothing about **A**, and knowing **A** is false tells us nothing about **B**. The meanings of the other connectives are immediately clear from their digraph lines. The graph is, of course, merely a way of displaying the structure of the connective's truth table.

Let us see how the system applies to an actual problem of the sort often found in elementary logic textbooks.

Problem 1. There are three women, Amy (A), Bertha (B), and Carol (C). We take "born in Texas" to be the equivalent of true, and "not born in Texas" to be the equivalent of false. We are given the following premises:

1. If Amy was born in Texas, then Bertha was born in Texas.

2. Either Bertha was born in Texas, or Carol was born in Texas, but not both.

3. Either Amy was born in Texas, or Carol was born in Texas, or both.

4. Bertha was born in Texas.

In the notation of the propositional calculus, the premises are:

1. $A \supset B$

2. $B \neq C$

3. $A \vee C$

4. B

We wish to learn, if possible, whether Amy and Carol were born in Texas. It is possible that the premises may harbor a contradiction in which case no conclusions can be reached. The combined premises also may leave open the question of where Amy was born, or Carol, or both.

Figure 4 (top) shows how the premises are digraphed. We simply take them in order. After the three graphs are drawn for the first three premises, and B is circled (on the basis of the fourth premise), all identical terms are joined by undirected lines to make a single connected graph.

We start exploring it at B because we know B is true. We cannot travel west along the line from the first circled B because the arrow prohibits it, but we can go from the second circled B to a $\sim$ C. We circle $\sim$ C as shown in Figure 4. From $\sim$ C we can travel to another $\sim$ C and circle it. From there, we can go along the directed line to A and circle it. From A we go to the leftmost A and circle it. The path leads back to B. The digraph now looks as shown. We see that all the As, Bs and $\sim$ Cs are affirmed. We conclude that Amy and Bertha were born in Texas, Carol was not. There is no way to

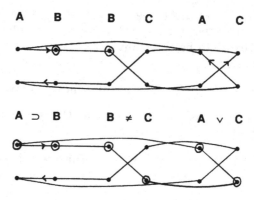

Figure 4. Solving Problem 1 with digraphs.

continue exploring the graph. No contradiction (affirming both a term and its negation) was encountered, therefore the problem is solved. If premise 4 had asserted that Bertha was not born in Texas ($\sim$ B), a similar exploration of the digraph would have affirmed $\sim$ A, $\sim$ B and C.

It is easy to see how the technique can be extended in a chain to take care of any number of binary premises, and involving any number of terms. There are techniques for using such digraphs to handle compound statements (statements with parentheses) but, as is the case also with Venn circles, the system becomes too complicated to be of much interest.

If the premises do not affirm the value of any single term, they may still permit determining the value of one or more terms. In such cases one must explore the digraph by making assumptions about values to find out if any assumptions can be eliminated by a contradiction. To see how this works consider the following problem.

Problem 2. A woman may or may not be Ann (A or $\sim$ A), may or may not be beautiful (B or $\sim$ B), and may or may not be clever (C or $\sim$ C). We are told:

1. The woman is either Ann or she is clever but not both.

2. She is not both beautiful and clever.

3. She is either Ann or beautiful, or both.

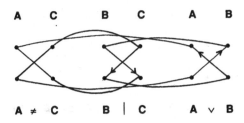

Figure 5. Solving Problem 2 with digraphs.

In notation:

1. $A \neq C$

2. $B \mid C$

3. $A \vee B$

What can be deduced about the woman:

The digraph is shown in Figure 5. To explore it, we start with any assumption we please. Let's assume the woman is Ann. We circle the two As. Traversing the lines allows us to circle two $\sim$ Cs. She is not clever. No contradiction is encountered, but we have learned nothing about whether Ann is beautiful. We try again, erasing the circles and circling $\sim A$ to see what follows if we assume the woman is not Ann. Following the lines quickly leads to contradictions—forcing us to affirm all the terms and their negations. We need go no further. The premises are consistent only with the assumption that the woman is Ann, who is not clever. Whether she is beautiful is undecidable.

If we assume that the woman is beautiful, we find that she must be Ann, and not clever. But we can also assume she is not beautiful, and hence find that she is Ann and not clever. If we assume she is clever, the digraph leads into contradictions. If she is not clever, the digraph shows she must be Ann, with her beauty undecided.

It goes without saying that, as with all good logic diagrams, the digraph provides a simple way to prove tautology. If two statements are identical, their digraphs will be identical. For example, De Morgan's well-known pair of laws asserts that $\sim (A \bullet B)$ (which is the same as $A \mid B$) is equivalent to $\sim A \vee \sim B$; and $\sim (A \vee B)$ is equivalent to $\sim A \bullet \sim B$. The equivalences are obvious when we digraph each side of either law and see that the digraphs are identical.

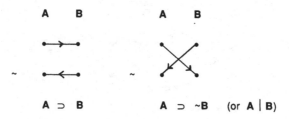

Figure 6. Negating a term.

To change the value of a single term X or $\sim X$ in a binary relation, imagine that the points at positions X and $\sim X$ are exchanged, carrying with them any lines that are attached. Think of a line as an elastic string, one end of which moves with a moving point while its other end remains fixed. Figure 6 shows, for example, $A \supset B$ is changed to $A \supset \sim B$.

To change a digraph of an entire binary relation to its negation, first change each undirected line to a double line with arrows going both ways. Take the front end of each line and move it to the term's other point—that is, move it either up or down. Finally, if the result is a double line, replace it with an undirected one. Figure 7 shows an example. We want to negate $A \supset B$. When the front ends of the two directed lines are switched, we get the middle digraph. The double line with arrows going both ways can now be replaced with an undirected line as shown on the right. We see that the negation of $A \supset B$ is $A \bullet B$. In words, if it is false to assert that A implies B, then A must be true and B must be false.

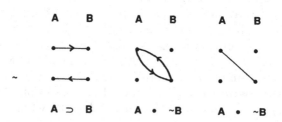

Figure 7. Negating a connective.

The logic digraphs can be modified and extended in many ways that suggest further study. What is the best way to digraph compound statements? Can digraphs handle class-inclusion logic with existential quantifiers such as the syllogisms premise "some A is B"? Can digraphs be applied to other logics such as the logic of strict implication and other modal logics? Can they handle multivalued logics by adding more points to each term and perhaps using different colors for the lines to indicate the possible values of the connectives? Can digraphs be applied to fuzzy logic?

The authors would be please to hear from anyone who investigates these questions, or who introduces digraphs in elementary logic courses and cares to report on how the students responded.

References

[1] Gardner, M., *Logic Machines and Diagrams*, revised edition, University of Chicago Press, Chicago, 1982.

[2] Venn, J., *Symbolic Logic*, Macmillan, London, New York, 1894.

[3] Harary, F., Norman, R., and Cantright, D., *Structural Models: an Introduction to the Theory of Directed Graphs*, Wiley, New York, 1965.

Chapter 5
Mathematics and Wordplay

Many mathematicians enjoy wordplay, and for obvious reasons: It is almost a branch of combinatorial mathematics. The pleasure derived from solving a combinatorial problem is very much like the pleasure of solving a cryptogram or a crossword puzzle, or constructing a good palindrome. Given the formal system of arithmetic, ancient mathematicians asked themselves whether the digits 1 through 9 could be placed in a three-by-three matrix so that rows, columns, and the two diagonals had the same sum. This is not much different from asking if, given the formal rules of English, one can construct a three-by-three word square in which each row, column, and main diagonal is a different word.

There is, of course, a difference between combinatorial mathematics and word play. Mathematics is embodied in the structure of the universe. Although mathematical systems are free inventions of human minds, they have astonishing applications to nature. No one expected non-Euclidian geometry to be useful, but it proved to be essential to relativity theory. Boolean algebra seemed useless until—surprise!—it turned out to model the electrical networks of computers. There are hundreds of other outstanding instances of what physicist Eugene Wigner has called the "unreasonable effectiveness of mathematics."

Think of the letters and words of a language, together with its rules, as a formal system. Although the words have arbitrary meanings assigned to them by minds, and there may be a "deep structure" of syntax that conforms to logic, the words themselves have no reality apart from a culture. Butterflies are all over the world, but you will not find the word "butterfly" by looking through a telescope or microscope. Once the word becomes attached to butterflies, however, it is amusing to note that butterflies flutter by. Because language, unlike

---

This article first appeared in *Word Ways* (February 1993).

mathematics, is "artificial," word play has more in common with, say, inventing card tricks or playing chess.

The combinatorial nature of word play is underscored by the recent use of computers for solving word problems. Disks containing all the words of a language are now available. With suitable programs they can be used to construct word squares, find anagrams, shortest word ladders, and so on. I wouldn't be surprised if some day computers solve complicated crosswords as easily as they now solve chess problems.

It is worth noting that, in both mathematics and world play, solving a problem is curiously like confirming a theory. In solving a cryptogram, for example, one first makes conjectures. Is a single-letter word A or I, or maybe O? Is ABCA the word "that"? Such conjectures are then tested to see if they lead to contradictions. If they lead to other words, they gain in their probability of being correct. Eventually a point is reached at which one is certain that a cryptogram has been cracked even though not all its letters are known.

One is tempted to say that when all words are known, one can be absolutely certain a cryptogram has been solved. This is not the case because, especially if the cryptogram is short, there could be another solution that the composer of the puzzle intended. If, however, the cryptogram is long, such an uncertainty becomes vanishingly small. This is true also in science. When there is a large abundance of facts explained by a theory, such as by the Copernican theory or the theory of evolution, certainty reaches a probability of $0.999999999\ldots$.

Now for a deep metaphysical question. If chess had not been invented, is there a sense in which theorems about chess can be said to exist? Assuming the formal system of chess, and given a certain position on the board, is it permissible to say that there *is* a mate in three moves even if no one has posed the problem? Assuming the structure of a deck of cards, is there a sense in which a good card trick is somehow "out there," in a Platonic realm of universals, even if no cards existed?

Suppose there were no English language. Would it be meaningful to say that given such a language, there is a sense in which a certain anagram "exists" even if no one spoke English? It is something like asking if a certain number with a million digits is prime or composite before anyone tested the number to find out. Well, not quite, because arithmetic certainly "exists" as a formal system. Anyway, most mathematicians are Platonists who believe that, no matter how bizarre, or how far removed from reality a system can be, they "discover" its theorems rather than invent them. Even though English is a human construction, nowhere to be found in nature, is there a sense in which its word plays are "real" before anyone finds them? I leave answering this question to my readers.

Postscript

I mentioned the task of constructing a three-by-three word square on which every row, column, and main diagonal is a common English word. In 1999 Donald Knuth, the noted computer scientist now retired from Stanford University, made a computer search for such squares. The program used a list of 604 three-letter words that are in common use, excluding proper names, acronyms, and abbreviations. It found 148 solutions. Here are two examples:

WHO   PEA
ION   URN
TEE   BAY

If duplicate letters are excluded, then such a square provides an amusing game isomorphic with tic-tac-toe. The square's nine letters are on nine cards. Players take turns drawing a card and the first to form one of eight words is the winner. Because tic-tac-toe is also a tie. However, if the word square is not shown, a person knowing the square, and knowing how to play a perfect game of tic-tac-toe, can win easily against opponents unaware of the isomorphism.

When Knuth checked his 148 word squares for solutions with no duplicate letters, he found none!

In a short article in *Word Ways* (August 1999) I asked the question of whether such a square exists. The article is reprinted here:

Tic-Tac-Toe Played as a Word Game

In problem 48 in *Your Move* (McGraw-Hill, 1971), David Silverman described a linguistic version of tic-tac-toe consisting of a stockpile of the words ARMY, CHAT, FISH, GIRL, HORN, KNIT, SOUP, SWAN, and VOTE. Players alternately select words, and the first to collect three words sharing a common letter is the winner.

In the single-letter analogue of Silverman's game, players draw alternately from a stockpile of nine different letters; the first to select the letters forming one of a specified list of nine words is the winner. To make it easy to remember these words, they can be written in the form of a three-by-three word square in which both diagonals are also words.

Squares are easy to find if you allow abbreviations, acronyms, proper names or foreign words. However, I'm half-convinced that there is no solution with common everyday words. It's almost spooky how you can find seven words but not the eighth. My near misses:

NOS	ARE	HOP	BED	FLU	DOS
EAT	SIN	EAR	OAR	AIR	EAT
WRY	POD	SKY	WHY	TEN	WRY

In the first square, the first word can be used in a sentence such as "The NOS have it" summarizing a vote. Change the N to L, and the W to D, and the first word, LOS, is the first half of Los Angeles. If given names are allowed, consider the second square with RIO. The third square uses the contraction HE'S; the sixth uses DOS, as in the phrase "dos and don'ts". There must be dozens or even hundreds of squares using words from, say, the Official Scrabble Players Dictionary, but these invariably employ less-familiar words than the ones in the squares above.

Ross Eckler generated an interesting set of squares that all use the French word EAU, found in English phrases like "eau de cologne" or "eau de vie":

FOG	HOG	LOG	LOB	FOB	GOB
EAU	EAU	EAU	EAU	EAU	EAU
DRY	PRY	TRY	TRY	DRY	TRY

Chapter 6
Steiner Trees on a
Checkerboard

Introduction

Suppose a finite set of n points are randomly scattered about in the plane. How can they be joined by a network of straight lines with the shortest possible total length? The solution to this problem has practical applications in the construction of a variety of network systems, such as roads, power lines, pipelines, and electrical circuits.

It is easy to see that the shortest network must be a tree, that is, a connected network containing no cycle. (A *cycle* is a closed path that allows one to travel along a connected path from a given point to itself without retracing any line.) If no new points can be added to the original set of points, the shortest network connecting them is called a *minimum spanning tree*.

A minimum spanning tree is not necessarily the shortest network spanning the original set of points. In most cases a shorter network can be found if one is allowed to add more points. For example, suppose you want to join three points that form the vertices of an equilateral triangle. Two sides of the triangle make up a minimum spanning tree. This spanning tree can be shortened by more than 13 percent by adding an extra point at the center and then making connections only between the center point and each corner (see Figure 1). Each angle at the center is 120°.

A less obvious example is the minimum network spanning the four vertices of a square. One might suppose one extra point in the center would give the minimum network, but it does not. The shortest network requires, in fact, *two* extra points (see Figure 2). Again all

---

This article, coauthored with Fan Chung and Ron Graham, first appeared in *Mathematics Magazine* (April 1989).

length 2   length $\sqrt{3}$

Figure 1.

angles around the extra points in the network are 120°. The network with one extra point in the center has length $2\sqrt{2}$, or about 2.828. The network with two extra points reduces the total length to $1 + \sqrt{3}$, or about 2.732.

One of the first mathematicians to investigate such networks was Jacob Steiner, an eminent Swiss geometer who died in 1863. The extraneous points that minimize the length of the network are now called Steiner points. It has been proved that all Steiner points are junctions of three lines forming three 120° angles. The shortest network, allowing Steiner points, is called a *minimum Steiner tree*. Minimum Steiner trees are almost always shorter than minimum spanning trees, but the reduction in length usually depends on the shape of the original spanning tree. It has been conjectured [9] that for any given set of points in the plane, the length of the minimum Steiner tree cannot be less than a factor of $\sqrt{3}/2$, or about 0.866, times the length of the minimum spanning tree; the result has been proved, however, only for three, four, and five points [10], [12].

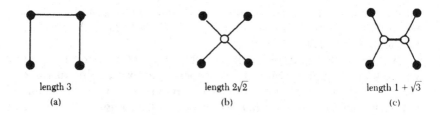

length 3   length $2\sqrt{2}$ length $1 + \sqrt{3}$
(a)   (b)   (c)

Figure 2.

Many properties of minimum Steiner trees can be found in the excellent (but somewhat out-of-date) survey paper of E. N. Gilbert and H. O. Pollak [9]. The best current lower bound for the ratio of the minimum Steiner tree to the minimum spanning tree is 0.8241... (see [4]).

There are many ways to construct a minimum spanning tree. One of the simplest methods is known as a greedy algorithm, because at each step it bites off the most desirable piece. First find two points that are as close together as any other two and join them. If more than one pair of points are equally close, choose any such pair. Repeat this procedure with the remaining points in such a way that joining a pair never completes a circuit. The final result is a spanning tree of minimum length. This algorithm is due to Kruskal in a 1956 paper [11].

Given the simplicity of Kruskal's greedy algorithm for the construction of minimum spanning trees, one might suppose there would be correspondingly simple algorithms for finding minimum Steiner trees. Unfortunately, however, this is almost certainly not the case. This task belongs to a special class of "hard" problems known in computer science as NP-complete problems. When the number of points in a network is small, say 10 to 20, there are known algorithms [5], [13] for finding minimum Steiner trees in a reasonably short time. As the number of points grows, however, the computing time needed increases at a rapidly accelerating pace. Even for a relatively small number of points the best algorithms currently available could take thousands or even millions of years to terminate. Most mathematicians believe no efficient algorithms exist for constructing minimum Steiner trees on arbitrary sets of points in the plane [7], [8].

Imagine, however, that the points are arranged in a regular lattice of unit squares, like the points at the corners of the cells of a checkerboard. Is there a "good" algorithm for finding a minimum Steiner tree spanning the points of such regular patterns? In particular, what is the length of the minimum Steiner tree that joins the 81 points at the corners of a standard checkerboard? Is the tree in Figure 3 the solution?

Many problems involving paths through points in the plane, which are difficult when the points are arbitrary, become trivial when the points from regular lattices. One might expect that the task of spanning points in such arrays by minimum Steiner trees would be equally trivial. On the contrary, this problem seems to be surprisingly elusive. Up to now, only minimum Steiner trees for 2 by n rectangular arrays

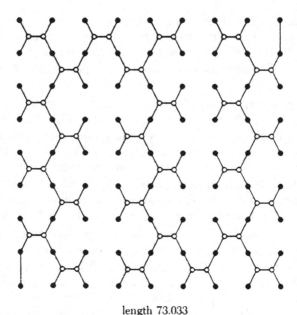

length 73.033

Figure 3.

of points have been constructed [3]. Aside from this special case, very little seems to be known about how to find minimum Steiner trees for rectangular arrays when the number of points on each side is greater than 2.

In this paper, we will summarize various problems, conjectures, and some partial results on the minimum Steiner trees for rectangular arrays. Section 2 contains the shortest known trees for square lattices of small size. These trees consist of copies of the symmetrical tree on four points (see Figure 2 (c)), which we call X from now on, together with a small number of "exceptional" pieces. For example, the conjectured solution for 64 points is a union of 21 X's (see Figure 5). In Section 3, we give a proof that a rectangular array can be spanned by a Steiner tree made up entirely of X's if and only if the array is a square and the order of the square is a power of 2. Further questions are proposed in Section 4.

Tree	Symbol	Length
	E	$e = 1$
	T	$t = \dfrac{1 + \sqrt{3}}{\sqrt{2}} = 1.93185\ldots$
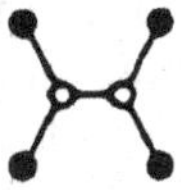	X	$x = 1 + \sqrt{3} = 2.73205\ldots$
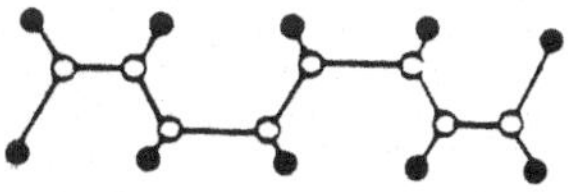	L	$l = \sqrt{35 + 20\sqrt{3}} = 8.34512\ldots$

Figure 4.

Short Steiner Trees on Square Lattices

In this section, we will first show the shortest Steiner trees we currently know for square lattices of size up to 14 by 14. We will then discuss a scheme for constructing Steiner trees for large square lattices form the small ones. Among all the constructions, only the patterns for the 2 × 2, 3 × 3 and 4 × 4 squares have been proved to be minimum Steiner trees (unpublished results of E. J. Cockayne). The constructions for square lattices of orders 2 to 9 were contained in the June 1986 issue of *Scientific American* [6]. The trees for square lattices of sizes 10 by 10 and 22 by 22 in the same article were soon improved by many readers. The current best tree for the 10 by 10 square lattice is due to one of the authors (RLG) and the best tree for the 22 by 22 square lattice is due to Eric Carlson [1]. His construction has the same total length as our general construction. Overall, the constructions fall naturally into six classes, depending on what n is modulo 6, with the rare (and remarkable) exceptions that occur when n is a power of 2. It seems that the Steiner trees for square lattices are always formed by attaching small minimum Steiner trees as an edge E (for the 1 by 2 array), X (for the 2 by 2 array), a triangle T (formed from three vertices of a square) and L (for the 2 by 5 array). We will use the notation given in Figure 4.

The tree L is a minimum Steiner tree on the regular 2 by 5 array [3]. The minimum Steiner tree for the 2 by n array with n even is just made up of X's joined together by edges. On the other hand the minimum Steiner tree for the 2 by n array with n odd has length $\frac{1}{2}((n(2 + \sqrt{3}) - 2)^2 + 1)^{1/2}$. The constructions for $n \times n$, $n \leqslant 14$, are illustrated in Figure 5.

n	Conjectured Minimum Steiner Tree	Length
2	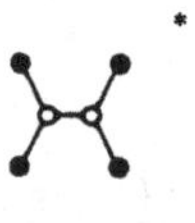	$x = 2.73205\ldots$
3	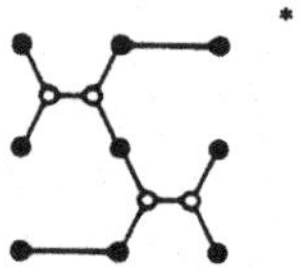	$2x + 2 = 7.46410\ldots$

*known to be optimal

Figure 5.

n	Conjectured Minimum Steiner Tree	Length
4		$5x = 13.66025\ldots$
5		$7x + 3 = 22.12436\ldots$
6		$11x + t = 31.98441\ldots$
7		$15x + 3 = 43.98076\ldots$

Figure 5 continued.

n	Conjectured Minimum Steiner Tree	Length

8

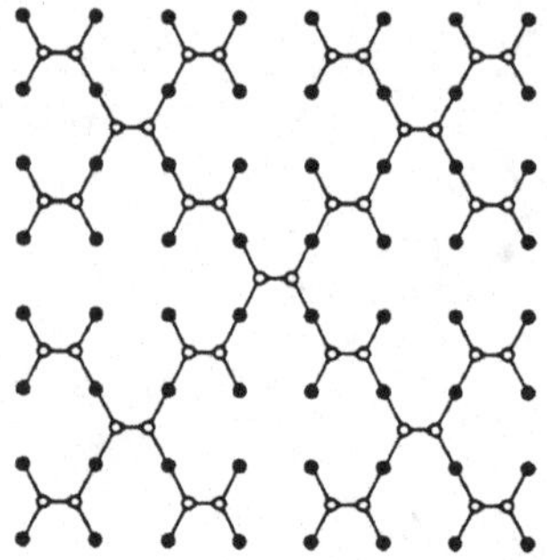

$21x = 57.373067\ldots$

9

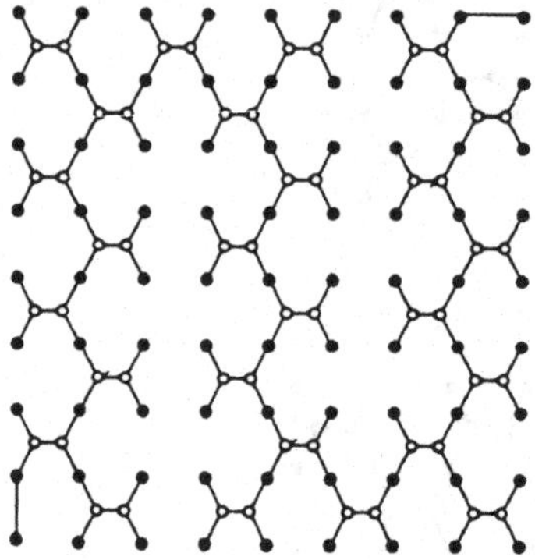

$26x + 2 = 73.03332\ldots$

10

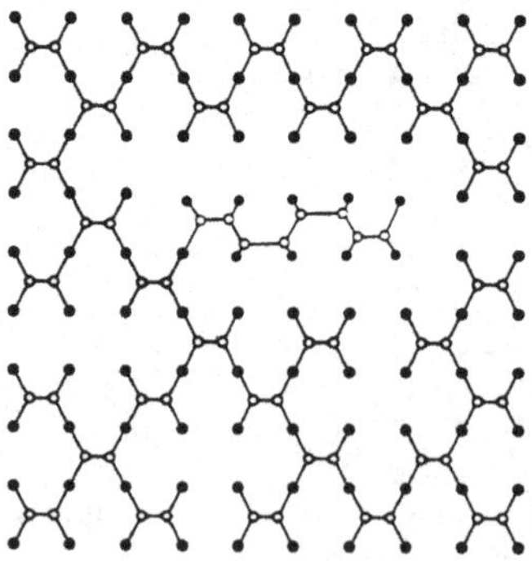

$30x + l = 90.30664\ldots$

Figure 5 continued.

n	Conjectured Minimum Steiner Tree	Length

11

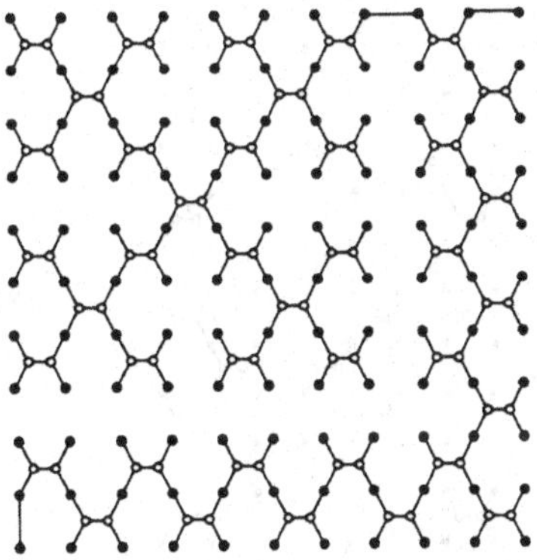

$39x + 3 = 109.54998$

12

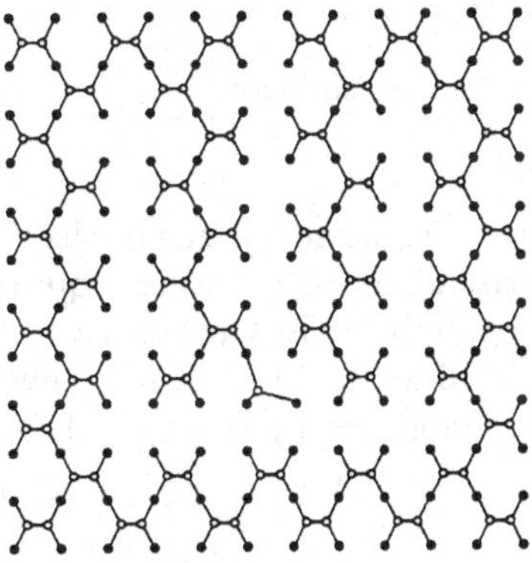

$47x + t = 130.33824\ldots$

13

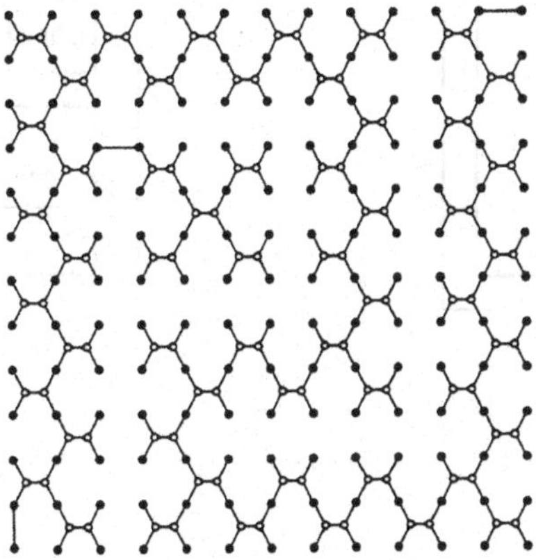

$55x + 3 = 153.26279$

Figure 5 continued.

n	Conjectured Minimum Steiner Tree	Length

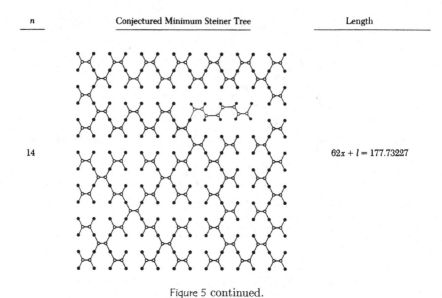

| 14 | | $62x + l = 177.73227$ |

Figure 5 continued.

To construct Steiner trees for large square lattices with orders not equal to a power of 2, we will always use a "core" square with a "folded band of width 3" wrapped around it in various ways. (The only core squares we need are 6×6, 10×10, 14×14 for n even, and 0×0, 4×4 and 8×8 for n odd.) The general pattern is shown in Figure 6.

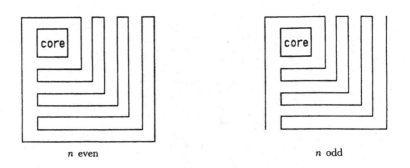

n even   n odd

Figure 6.

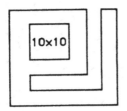

Figure 7.

Each additional "fold" of the strip adds 6 to the size of the grid.

For example, for $n = 22$, we see that $22 \equiv 10(\text{mod } 6)$ so we use a 10 × 10 core with 3 (doubled) folds of the band as shown in Figure 7.

Of course, the strip must be broken and connected to the core. (See the detailed picture in Figure 8. We note that in order to make the connection to the strip, the corresponding place in the core must be "striplike".) When n is $6k$, the core is 0 × 0, i.e., empty, so we don't have to connect it to the band. In this case, we only have to break the band and reconnect the two isolated points with a T.

For *odd* n, the band doesn't form a cycle but is open at each end, leaving two isolated points at the end. You can see this happening in the conjectured minimum Steiner trees for 9 × 9 and 13 × 13. In this case, when $n \not\equiv 0(\text{mod } 6)$ we only need an E to connect the core to the band. Summarizing these results, we have for $n \geqslant 15, n \neq 2^t$:

n	Length of conjectured minimum Steiner tree for G_n
$6k$	$(12k^2 - x + t)$
$6k + 1$	$(12k^2 + 4k - 1)x + 3$
$6k + 2$	$(12k^2 + 8k - 2)x + l$
$6k + 3$	$(12k^2 + 12k + 2)x + 2$
$6k + 4$	$(12k^2 + 16k + 2)x + l$
$6k + 5$	$(12k^2 + 20k + 7)x + 3$

where $x = 1 + \sqrt{3}, l = \sqrt{35 + 20\sqrt{3}}$, and $t = (1 + \sqrt{3})/\sqrt{2}$. Of course, for $n = 2^t$, any right-thinking person would guess that the length of the minimum Steiner tree for G_{2^t} is just $(\frac{1}{3})(4^t - 1)x$ but unfortunately we can't even prove this for $t = 3$!

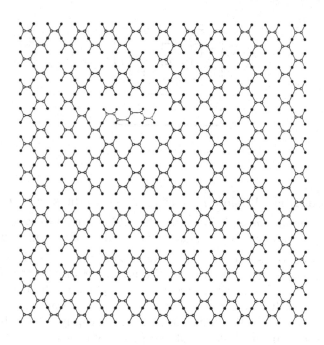

Figure 8.

Square Lattices for Powers of 2

Here we will give the proof of the main result in this paper.

Theorem
 If a rectangular array can be spanned by a Steiner tree made up entirely of X's, then the array is a square of size 2^t by 2^t for some $t \geqslant 1$.

Proof. We start by giving each 2×2 "cell" a pair of coordinates (i, j) in the obvious way, where the lower left-hand cell has coordinates $(0, 0)$.

Let's call a cell (i, j) *even* if $i + j$ is even. Also, let's call a cell *occupied* if it has an X in it. *Suppose* our $a \times b$ array has an X-tree, that is, a Steiner tree formed by X's.

Fact 1. Only *even* cells can be occupied.

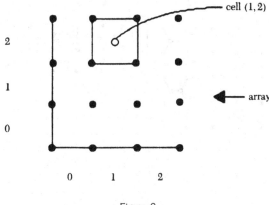

Figure 9.

Proof. The occupied cells must be connected. Furthermore, there can't be two *adjacent* occupied cells. Thus, occupied cells can only touch each other diagonally (as shown), in which case they are both even or both odd. However, the corner cell $(0,0)$ is occupied and even. Thus, *all* occupied cells are even.

As a consequence, we see that a and b must both be even. Call an even cell (i,j) *doubly even* if i and j are both even. Otherwise, call it *doubly odd*.

Fact 2. Every doubly even cell must be occupied.

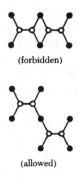

(forbidden)

(allowed)

Figure 10.

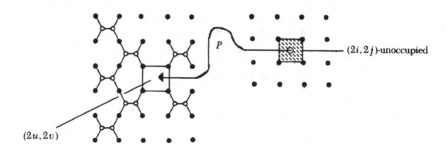

Figure 11.

Proof. First note that *all* the even cells on the boundary must be occupied and are doubly even. Suppose we had some interior doubly even cell $(2i, 2j)$ that was not occupied. Then, we must be able to draw a path P from the center of $(2i, 2j)$ that goes to the outside of the array and doesn't pass through any occupied cell. (This is because the complement of our X-tree must be a connected set.) But if any even cell is *unoccupied* then *all four* of its even neighbors must be *occupied* (since, otherwise, one of its corner points would be disconnected.) This now implies that the only even cells P can pass through are doubly even ones, and never a doubly odd one.

Now, focus on a doubly even cell $(2u, 2v)$ just inside the boundary that P tries to pass through on its way to the outside. Since $(2u, 2v)$ must be unoccupied (because P is going through it), all of its neighbors must be occupied. In particular, this forms a *barrier* with the adjacent occupied boundary cells that prevents P from going through to the outside here. But this happens wherever P tries to reach the outside since all the even cells on the boundary are occupied. Thus, P can never reach the outside, which is a contradiction. Hence, the hypothesis that there is an unoccupied doubly even cell is untenable, and the assertion is proved.

Therefore, in order to know what our X-tree is, we only have to know which (additional) *doubly odd* cells are occupied. Look at the picture for an 8×8 array in Figure 12.

Next to the 8×8 array, a 4×4 array is drawn. Notice that there is natural correspondence between the array of 9 doubly odd

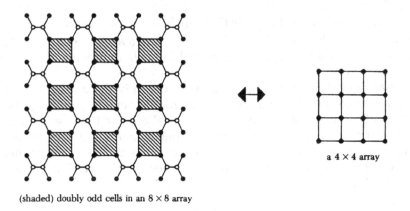

(shaded) doubly odd cells in an 8 × 8 array

a 4 × 4 array

Figure 12.

cells (shaded) in the 8 × 8 and the array of 9 cells in the 4 × 4. The key observation now, which is not hard to check, is that *the set of occupied doubly odd cells in the 8 × 8 must correspond exactly to an X-tree* in the 4 × 4 array. Namely, consecutive doubly odd cells in the same row or column cannot both be occupied (or we get a cycle), and all points in the array must be joined together.

In effect, the doubly odd cells on the 8 × 8 array form a "stretched-out" version of all the cells of a 4 × 4 array.

More generally, this argument shows that if an $a \times b$ array has an X-tree then $a = 2A, b = 2B$, and, furthermore, the smaller $A \times B$ array must also have an X-tree.

We now can apply this repeatedly (similar to Fermat's method of infinite descent, except that we stop at 2 × 2) to get the conclusion that only $2^t \times 2^t$ arrays can have X-trees. This completes the proof of the theorem.

Concluding Remarks

Of course, the main open problem is to determine the minimum Steiner trees for all (or even infinitely many) square lattices. It is embarrassing that even for 2^t by 2^t arrays, we still can't prove optimality for the "obviously" correct X-tree.

If we assume the distance between adjacent vertices in the lattice is 2, then it is not hard to show that any Steiner tree (minimum or not) has length of the form $\sqrt{a + b\sqrt{3}}$ where a and b are integers (possibly negative). Conceivably, this fact could be of help in proving optimality in some cases.

An interesting related question is to use only the 2×2 minimum Steiner trees and the smallest possible number of single edges (called E's) to form a spanning tree of $G(m, n)$. We know that for m and n both large enough, only a bounded number of E's are ever needed. When m is small however (where we can assume $m \leqslant n$), we may need arbitrarily many E's. Some examples of this are illustrated in Figure 13.

Acknowledgements

The authors would like to acknowledge the meticulous help of Nancy Davidson in preparing the figures for this paper.

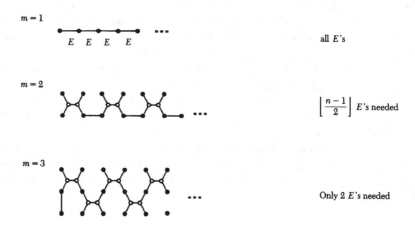

Figure 13.

$m = 4$

$\left\lfloor \dfrac{n-1}{4} \right\rfloor$ E's needed

$m = 5$

A *bounded* number
of E's needed
(at most 4)

$m = 6$

At most
2 E's needed

$m = 7$

At most
4 E's needed

Figure 13 continued.

References

[1] E. Carlson, private communication.

[2] F. R. K. Chung and R. L. Graham, "Steiner trees for the regular simplex," *Bull. Inst. Math. Acad. Sinica* 4 (1976), 313–325.

[3] F. R. K. Chung and R. L. Graham, "Steiner trees for ladders," *Annals of Discrete Math.* 2 (19780, 173–200.

[4] F. R. K. Chung and R. L. Graham, "A new bound for Euclidean Steiner minimal trees," *Annals of the N.Y. Acad. Sci.* 440 (1985), 328–346.

[5] E. J. Cockayne and D. E. Hewgill, "Exact computation of Steiner minimal trees in the plane," *Inform. Proc. Letters* 22 (1986), 151–156.

[6] M. Gardner, "Mathematical games," *Scientific American* (June 1986), 16–22.

[7] M. R. Garey, R. L. Graham and D. S. Johnson, "The complexity of computing Steiner minimal trees," *SIAM J. Appl. Math* 32(1977), 835–859.

[8] M. R. Garey and D. S. Johnson, *Computers and Intractability, a Guide to the Theory of NP-Completeness*, W. H. Freeman and Co., San Francisco, CA, 1979.

[9] E. N. Gilbert and H. O. Pollak, "Steiner minimal trees," *SIAM J. Appl. Math.* 16 (1968), 1–29.

[10] R. Kallman, "On a conjecture of Gilbert and Pollak on minimal trees," *Stud. Appl. Math.* 52 (1973),141–151.

[11] J. B. Kruskal, "On the shortest spanning subtree of a graph and the traveling salesman problem," *Proc. Amer. Math, Soc.* 7(1956), 48-50.

[12] H. O. Pollak, "Some remarks on the Steiner problem," *J. Combin. Theory Ser.* A 24(3) (1978), 278–295.

[13] P. Winter, "An algorithm for the Steiner problem in the Euclidean plane," *Networks* 15 (1985), 323–345.

Postscript

Before the foregoing paper was published, I had devoted the last of my "Mathematical Games" columns in *Scientific American* (June 1986) to Steiner trees. When an updated reprint of this column appeared in

Last Recreations (Springer-Verlag 1997) I added an addendum and a longer bibliography. What follows here is that addendum and the list of references.

Two major breakthroughs relating to MSTs (minimal Steiner trees) have occurred since this chapter was written in 1986. In 1968 two Bell Labs mathematicians, H. O. Pollok and E. N. Gilbert, conjectured that the ratio of the length of an MST to the length of the minimal spanning tree for the same set of points is at least $\sqrt{3}/2 = 0.866...$, a savings in length of about 13.4%. This is the ratio for the two kinds of trees that join the corners of an equilateral triangle. (See the 1968 paper by Gilbert and Pollok.) In 1985 Ronald Graham and his wife Fan Chung raised the lower bound of the ratio to 0.8241. The proof was so horrible, Graham said, that he urged those interested *not* to look up their paper.

The problem was important enough to Bell Labs, where finding shorter networks is an obvious cost saving, for Graham to offer $500 to anyone who could prove the $\sqrt{3}/2$ conjecture. The prize was won in 1990 by two Chinese mathematicians, Ding Zhu Du, then a postgraduate student at Princeton University, and Frank Hwang, of Bell Labs. (See their 1992 paper.)

A simplex is a regular polyhedron, in any dimension, with a minimum number of sides, such as the 3-space tetrahedron. MSTs are known only for simplexes through five dimensions. (See the 1976 paper by Chung and Gilbert.) Calculating them for higher dimensions is far from solved. The MST for the corners of a unit cube is shown in Figure 14. Its length is 6.196....

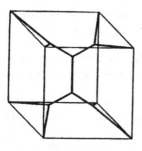

Figure 14.

Hwang and Du, in their 1991 paper, study MSTs on isometric (equilateral triangle) lattice points.

The other breakthrough, by five Australian mathematicians, was a complete solution to finding MSTs for both square and rectangular lattices of points in a matrix of unit squares. (See the 1995 Research Report by M. Brazil and his four associates.) In a 1996 paper by Brazil and five associates they confirmed the unpublished proof by Graham and Chung about the form of MSTs for sets of points at the vertices of a $2^k \times 2^k$ square lattice.

There is a growing literature on minimal *rectilinear* Steiner trees—trees with only horizontal and vertical lines. They have important applications to electrical circuit design. See Dana Richard's "Fast Heuristic Algorithms for Rectilinear Steiner Trees," in *Algorithmica*, Vol. 4, 191–207, 1989.

In a lecture on Steiner trees by Graham, which I had the pleasure of attending, he included the following points:

Jacob Steiner made no contributions to the theory of Steiner trees except to get his name attached to them. The points were earlier called Fermat points, but their existence was known even before Fermat's time.

The first proof that finding Steiner trees for n points is NP-complete was in the 1977 paper by Graham and two Bell Labs colleagues, Michael Garey and David Johnson. Also NP-complete is the problem of calculating the exact length of a minimal spanning tree. Intuitively it seems as if the greedy algorithm would make this easy. It is not easy because the points may not be at integer coordinates on the plane. As such points increase in number, calculating the exact length of the tree rapidly becomes more difficult.

References

[1] E. N. Gilbert and H. O. Pollok, "Steiner minimal trees," *SIAM Journal of Applied Mathematics*, Vol. 16(1), pp. 1–29, January 1968.

[2] Fan Chung and E. N. Gilbert, "Steiner trees for the regular simplexes," *Bulletin of the Institute of Mathematics Academy Sinica*, Vol. 4, pp. 313–325, 1976.

[3] F. H. Hwang, "On Steiner minimal trees with rectilinear distance," *SIAM Journal of Applied Mathematics*, Vol. 30, pp. 104–114, 1976.

[4] M. R. Garey, R. L. Graham, and D. S. Johnson, "The complexity of computing Steiner minimal trees," *SIAM Journal of Applied Mathematics*, Vol. 32, pp. 835–859, 1977.

[5] F. R. K. Chung and R. L. Graham, "Steiner trees for ladders," *Annals of Discrete Mathematics*, Vol. 2, pp. 173–200, 1978.

[6] Dale T. Hoffman, "Smart soap bubbles can do calculus," *The Mathematics Teacher*, Vol. 72(5), pp. 377–385, 389, May 1979.

[7] F. Chung and R. Graham, "A new bound for Euclidean Steiner minimal trees," *Annals of the New York Academy of Sciences*, Vol. 440, pp. 328–346, 1985.

[8] M. Bern and R. Graham, "The shortest-network problem," *Scientific American*, Vol. 260, pp. 84–89, 1989.

[9] F. Chung, M. Gardner, and R. Graham, "Steiner trees on a checkerboard," *Mathematics Magazine*, Vol. 62, pp. 83–96, April 1989.

[10] F. K. Hwang and D. Z. Du, "Steiner minimal trees on chinese checkerboards," *Mathematics Magazine*, Vol. 64, pp. 332–339, Decemer 1991.

[11] F. K. Hwang, D. S. Richards, and P. Winter, "The Steiner tree problem," *Annals of Discrete Mathematics*, Vol. 53, Amsterdam, 1992.

[12] R. B. Cohen, "Optimal Steiner points," *Mathematics Magazine*, Vol. 65, pp. 323–329, December 1992.

[13] D. Z. Du and F. K. Hwang, "A proof of the Gilbert-Pollok conjecture on the Steiner ration," *Algorithmica*, Vol. 7, pp. 121–135, 1992.

[14] R. Bridges, "Minimal Steiner trees for the three-dimensional networks," *The Mathematical Gazette*, Vol. 78, pp. 157–162, July 1994.

[15] M. Brazil, J. H. Rubinstein, J. F. Weng, N. C. Wormald, and D. A. Thomas, "Full minimal Steiner trees on lattice sets," *Research Report 14*, Department of Electrical Engineering, University of Melbourne, Australia, pp. 1–40, 1995.

[16] M. Brazil, J. H. Rubinstein, D. A.Thomas, J. R. Weng, and N. C. Wormald, "Minimal Steiner trees for rectangular arrays of lattice points," *Research Report 24*, University of Melbourne, Australia, pp. 1–28, 1995.

[17] M. Brazil, T. Cole, J. H. Rubinstein, D. A. Thomas, J. F. Weng, and N. C. Wormald, "Minimal Steiner trees for $2^k \times 2^k$ square laticces," *Journal of Combinatorial Theory*, Series A., Vol. 73, pp. 91–109, January 1996.

Chapter 7
Tiling the Bent Tromino

The monomino (unit square), domino, and straight tromino can each be cut into n congruent parts, for any integer n. This is obvious because each is a rectangle, and a rectangle can be cut into n congruent, parallel strips. The bent tromino (or L-tromino) is, therefore, the simplest polyomino that raises unanswered questions about the values of n that allow for perfect tiling.

F. Göbel askes for what values of n the bent tromino can be cut into n congruent parts [1]. He reproduces three dissections, in 2, 3, and 4 congruent parts, shown here in Figure 1, and asks if the tromino can be cut into five congruent parts.

Because a square can be cut into n congruent strips, it follows at once from the $n = 3$ case that perfect solutions exist for any n that is a multiple of 3. The $n = 4$ case solves an ancient puzzle usually given in terms of a farm that is to be divided into four congruent regions. The unique solution uses tiles of the same shape as the farm. In S. W. Golomb's terminology, such shapes are called rep-tiles.

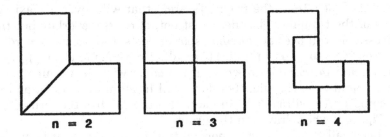

n = 2   n = 3   n = 4

Figure 1.

This article first appeared in the *Journal of Recreational Mathematics*, Vol. 22(3), 1990, pp. 185–191.

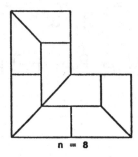

n = 8

Figure 2.

Because each of these rep-tiles can similarly be cut into four smaller replicas, it follows that perfect solutions exist for any n that is a power of 4. All three tilings in Figure 1 are surely unique, though I know of no proofs.

The tile for the $n = 2$ case is another rep-tile. Figure 2 shows how it can be cut into four smaller replicas to provide a solution for $n = 8$, and for solutions when n takes values in the series 8, 32, 128, 512,.....

Göbel also states that it is not hard to find solutions when n has the form $3m$ or m^2 or $2m^2$, and asks if there are perfect solutions for any n not of these three forms [1]. Solutions given here for $n = 10$ and 14 answer this question affirmatively. No solutions are known when n is prime greater than 3, nor are impossibility proofs known for such cases.

If a perfect tiling is impossible, we can seek for a tiling by congruent tiles, each $1/n$th the tromino's area, that will cover a maximum area of the tromino. The tiles must not, of course, overlap, but they necessarily will overlap, at some spot or spots, the tromino's border.

Figures 3 to 8 show the "best" results I have found for $n = 5$, 7, 10, 11, 13, and 14. Each is drawn on graph paper with the tromino's short sides equal to n The tile, therefore, will have an area of $3n$, and the tromino's area will be $3n^2$. In each case the tile has the form shown in Figure 9 when drawn on the graph.

Let's call Figure 9 the "canonical tile" for a given n. It is my first conjecture that for every n a "best" solution can be achieved by canonical tiling. When n is a multiple of 3, the simplest tiling is, of course, by rectangles of size $3 \times n$, but in every case the tromino can also be tiled canonically.

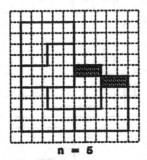

Figure 3.

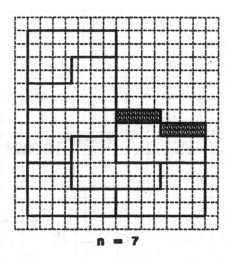

Figure 4.

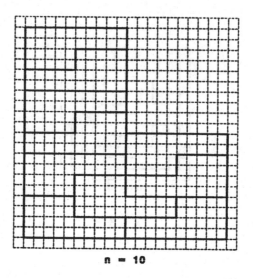

n = 10

Figure 5.

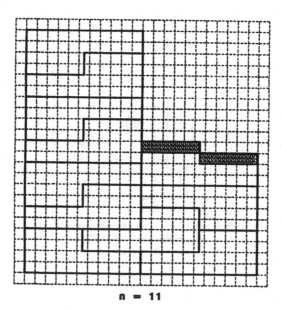

n = 11

Figure 6.

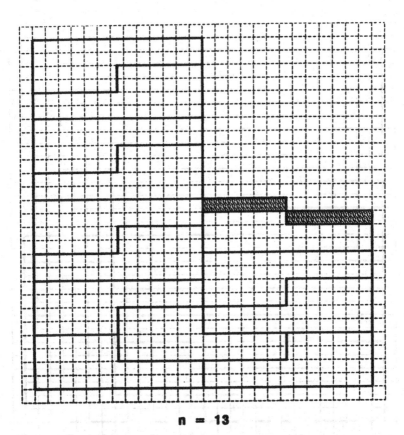

n = 13

Figure 7.

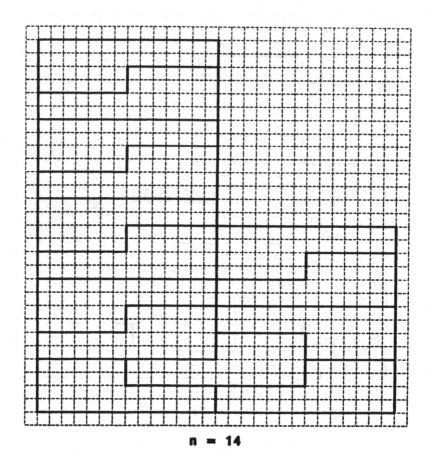

n = 14

Figure 8.

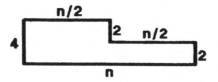

Figure 9.

These patterns, and similar patterns for higher n that are easily found, suggest the following additional conjectures:

1. Perfect canonical tilings exist for every n greater than 2.

2. There are no perfect canonical tilings for any odd n greater than 3 that is not a multiple of 3. This set of course includes all primes above 3.

3. For every n that has no perfect tiling, a canonical tiling will cover all but $\frac{1}{6}n$ of the total area, and no "better" tiling is possible.

To obtain a canonical tiling, the procedure is simple. Tile down the tromino's left side with $6 \times n$ rectangles, each of which can be cut into two canonical tiles. When this column of rectangles cannot be continued, tiles go at the bottom in a pattern that varies with n.

After the bottom tiles are suitably placed, continue upward on the right. If n is odd and not a multiple of 3, the topmost tile on the right will project above the border with a $1 \times \frac{n}{2}$ rectangle. This necessarily leaves uncovered a rectangle of the same size and shape. As n increases, the uncovered space becomes a smaller and smaller fraction of the total area, approaching but never reaching zero as the limit.

Canonical tiling clearly will not tile perfectly when n is odd and not a multiple of 3. The lowest composite case is 25. However, as Göbel recognized, square n's can be perfectly tiled with bent trominoes. Figure 10 shows tiling for squares of 3 through 7. Because each tromino can be cut into congruent halves (see Figure 1), obviously any square n tiling will provide a tiling for $2n$.

Nothing in this article rules out perfect tilings when n is odd, not square, and not divisible by 3. I would welcome any proofs or counterexamples to my conjectures.

References

[1] F. Göbel, Problem 1771—The L-Shape Dissection Problem, *Journal of Recreational Mathematics*, Vol. 22(1), pp. 64–65, 1990.

Postscript

Michael Beeler, in the *Journal of Recreational Mathematics* (Vol. 24, No. 1, 1992, pp. 65–69), "inverted" my approach to tiling L-trominoes with small stretched L-trominoes. Instead, he tiles horizontally stretched L-trominoes with normal bent trominoes, then stretches the

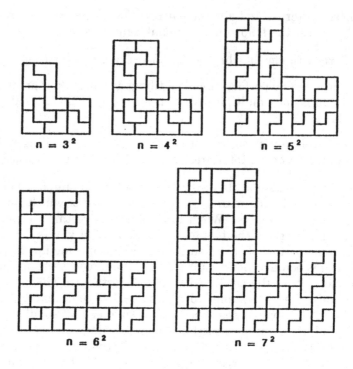

Figure 10.

structure vertically to make a normal bent tromino that is tiled by the stretched ones. Using this technique, he proves that for any composite n the bent tromino can be tiled with stretched ones. Left open is the question of whether a bent tromino can be dissected into a prime number of congruent shapes greater than 3.

Beeler also shows that if n is a square, a symmetrical tiling by bent trominoes is always possible. Figure 11 shows how this is done when $n = 9^2$.

Let k be the number of cells on a short side of a bent tromino, as shown in Figure 12. If k is not a multiple of 3, the bent tromino cannot be tiled with straight trominoes. The diagonal coloring insures that every straight tromino placed on the pattern must cover three cells that contain all three colors. But the number of cells of each color are *not* equal. They are in arithmetical progression. In the case pictured ($k = 4$), there are 15 gray cells, 16 black, and 17 white.

Figure 11. Beeler's symmetrical tiling when $n = 9^2$.

The diagonal 3-coloring provides a pleasant companion to the old brain teaser about whether a chess board, with diagonally opposite corner cells removed, can be covered with 31 nonoverlapping dominoes. Remove just one corner cell. Can the 63 remaining unit squares be covered with 21 straight trominoes? The diagonal coloring proves impossibility.

Figure 12. A three-color proof that the bent tromino with a short side of 4 cannot be tiled with straight trominoes.

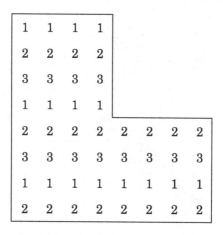

Figure 13. Another impossibility proof.

When I sent my coloring proof to Donald Knuth he realized at once that it could prove much more. He showed that if the m squares of a non-straight m-onomo are divided into $n \times n$ unit squares, the m-onomo cannot be tiled with n^2 straight m-onimoes when n is relatively prime to m. Thus a non-straight pentomino, its five squares divided into $n \times n$ unit squares, can be tiled with n^2 straight pentominoes if and only if n is a multiple of 5. A 6×6 square can't be packed with 9 straight tetrominoes ($m = 4$, which is relatively prime to $n = 3$). And there are endless other examples.

These results can also be obtained by labeling the rows of a polyomino 1, 2, 3, 1, 2, 3, ... Figure 13 shows how this applies to the L-tromino to prove it cannot be tiled with straight trominoes unless its short side is a multiple of 3. Any straight tromino placed on the pattern will cover three cells with a sum that is a multiple of 3. Hence if the pattern can be covered without overlapping, the total of all its cells must be a multiple of 3. In this case the total, 92, is not a mutliple of 3.

The main unsolved problem about L-tromino tiling, which may be very difficult, is to show that it can't be tiled with five congruent shapes. Perhaps this will fall out of a more general proof that tiling is impossible if the tromino's short side is a prime greater than 3.

Cutting the bent tromino into four congruent parts is a puzzle found in many early puzzle books. This page is from George Carlson's *Peter Puzzlemaker*, a 1922 book for children.

BAKER baked a delicious raisin cake and sold a quarter of it. The cake was a square, and the piece cut from it was also a square. Four hungry kids enter the shop to buy the rest of the cake. The baker wants three dollars for it, but the kids have only two dollars. As it is the youngest boy's birthday, the baker will let them have the cake for two dollars if they can tell him how to cut it into four pieces of exactly the same size and shape. How will they get the cake?

Figure 14.

Chapter 8

Covering a Cube with Congruent Polygons

* * *

Can the surface of a cube be covered, without overlap, with n congruent polygons, where n is any integer greater than 1?

* * *

The problem is unsolved.

Think of the polygons as pieces of paper that can be pasted on the surface of a cube, folding them when necessary. Althernatively, we can think of coloring the cube with n colors so that the regions can be unfolded to form n congruent polygons. The case of $n = 2$ is easily solved by two rectangles of size 3×1 for the unit cube. Because each rectangle can be divided into n congruent rectangles, it follows that the original problem can be solved for *any even* n.

The case of $n = 3$ is also easily solved by three 2×1 rectangles. Again, each rectangle can be cut into n congruent rectangles, therefore the original task is solved for *any* n *that is a multiple of 3*.

We now inquire whether there is a solution for $n = 5$. The answer is *yes*. The cube can be covered by five congruent *Latin crosses*, as shown in Figure 1.

This problem was posed in *Mathematics Magazine*, January 1976, by Veit Elser of San Jose, California. A solution by the proposer and by Michael Goldberg of Washington, D. C., appeared in *Mathematics Magazine*, Vol. 50 (1977), pp. 168–179.

The next higher n that remains unsolved in $n = 7$. *Can the cube be covered with seven congruent polygons?* This question is unanswered.

The above noted appeared in *Cubism for Fun* (No. 25, December 1990, Part 1, pp. 13).

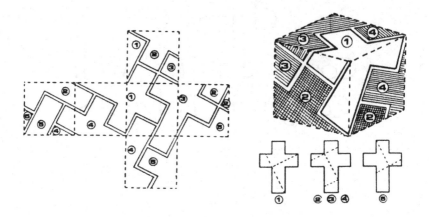

Figure 1. A cube covered by five congruent Latin Crosses. The folds are indicated by the dotted lines.

I here offer U.S. $50 to the first person who either finds a solution or proves it impossible.

Postscript

To my vast surprise Anneke Treep not only won my fifty dollars, she found four different strips that would wrap around a cube, each of which could be divided into *n* congruent parts! They are shown in Figure 2. Figure 3 shows how the first strip is easily cut into n congruent parts. Treep explained this in her article "Covering a Cube," in *Cubism for Fun* (No. 27, December 1991, pp. 16–17).

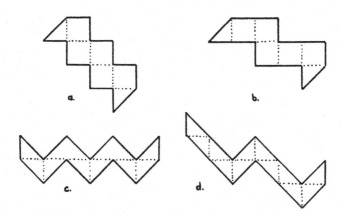

a.

b.

c.

d.

Figure 2.

Figure 3.

Chapter 9
Magic Tricks on a Computer

Several years ago I was involved in a software project involving a diskette that would play the role of a magician, performing magic tricks, some of which would mystify even a person running the programs. Nothing came of this project, but in planning material for the diskette (conjuring is my principal hobby) I accumulated an extensive file of tricks that could easily be programmed. Some magicians have thought along similar lines, and programs have been published in magic periodicals, but the field remains wide open to exploration by magicians in collaboration with clever programmers. Here are four typical ideas.

A Card Prediction

The trick begins with the computer throwing on the screen a picture of the back of a playing card. It remains there until the trick's climax.

You are told (in a booklet accompanying the diskette) to shuffle a deck of cards thoroughly, then type into the computer the names of the ten cards, in sequence, that are at the bottom of the shuffled deck. Leave the ten cards in the same position in the deck.

Holding the deck face down, start dealing cards from it face up to form a pile. As you deal, count backward from 10 to 1. All face cards are assigned a value of 10. If a card's value matches a called number, stop dealing and start a second pile. For example, assume you count 10, 9, 8, 7, 6, and a six is dealt on the count of six. Start a second pile, leaving the first pile with the six face up on top. This value obviously was determined by pure chance.

Three more piles are formed in exactly the same way. If no coincidence of card and number occurs, "kill" the pile by putting the next

This article first appeared in *Algorithm* (Jan/Feb 1990).

card face down on top of it. If all four piles are killed (this seldom happens), then shuffle the four piles together, replace the cards on top of the undealt cards, and start over.

After the four piles are completed, add the values of the cards on top of the unkilled piles. For example, one pile may have been killed, and a three, queen, and nine are face up on the other three piles. The queen counts as 10, so the sum is 3 + 10 + 9 = 22. Count to the 22nd card in the remainder of the deck and note its face.

A computer key, specified in advance (it is always the same key), is now pressed. Slowly the card on the screen rotates to disclose its other side. It will be the very card you seemingly chose at random!

How does the computer know? When you type the names of the ten cards at the bottom of the shuffled deck, into the computer, it ignores all the names except the ninth. This is the card whose face the computer puts on the screen as the face of the rotating card.

A bit of elementary algebra will convince you that the ninth card from the deck's bottom is always the one selected. The trick's most amusing aspect is that spectators have a strong (but completely false) impression that somehow the computer was able to put the chosen card on the screen, its face hidden, before the deck was shuffled.

Guessing A Word

In this trick you assist the computer in a sneaky way. The trick is intended for a group of onlookers who do not suspect that you play a role in its working.

Someone freely names a four-letter word. The computer then asks that person a series of four questions. Onlookers watch over your shoulder as you type his or her replies onto the screen.

The program can hold a list of, say, a hundred questions so that the four it asks are different each time the trick is repeated. The questions are totally irrelevant. For example, the first question could be: "In what city and state were you born?"

The next question could be: "What is the first line of one of your favorite poems?" The next two might be: "Who played the male lead in Casablanca?" and "What's the sum of ten thousand and fifteen thousand?" In each case you type exactly what the person tells you.

When a certain key is pressed, the chosen word flashes on the screen.

You must, of course, secretly signal the word. There are many ways to do this. Here is one of the simplest. The computer is programmed not to show on the screen the letter you type after hitting the space bar for the first time. Suppose the chosen word is "pink." When you type the answer to the first question, start the second word with p, then without hesitation go on to the letters of the second word. The screen will not show p or move the cursor. Spectators will be watching the screen, not your fingers. In the same way you give the computer i, n, k as you type answers to the last three questions.

The first published trick of this type, in which you secretly cue the computer while you type answers, was invented by Chris Morgan, former editor of *Byte*. In Morgan's version, the identity of a chosen card is conveyed by hitting or not hitting the space bar after you finish typing the answers to yes or no questions. You'll find this excellent trick described by Kee Dewdney in his *Scientific American* column for August 1986, and by Morgan in *Byte*, Fall 1979. Morgan provides a complete program in BASIC.

A Hat Trick

The screen shows a row of, say, three hats, brims down, and all alike. They are labeled ABC. Pressing a corresponding letter raises a hat a few inches. A second pressing lowers it. Beside the hats is a small bunny.

Stand on the far side of the room, back to the audience, while someone raises a hat, uses the mouse to put the rabbit under it, then lowers the hat over the rabbit. You turn around, return to the computer, and at once press the key that lifts the hat that hides the bunny. How do you know which hat to raise?

In the previous trick you secretly cue the computer. In this one, it secretly cues you. While the hat is being raised, the computer imperceptibly alters the picture of the hat one step to the right. (Assume ABC to be cyclic.) This can be done by placing a tiny dot somewhere on the hat, or by removing a tiny dot, or any other small change too small to be noticeable unless you know where to look. On the off chance that a person is closely watching the hat being raised, the change is made on another hat.

The trick obviously has many variants: a row of walnut shells and a pea; a row of inverted cups and a small object such as a ring, marble,

Your victim hides the rabbit under a hat.

Figure 1.

or match folder; a row of envelopes and a thousand-dollar bill; a row of doors behind which a leprechaun hides; and so on.

Guessing A Digit

Ask anyone for a dollar bill, and call attention to its eight-digit number. Read these digits aloud slowly. As you do speak, ask the bill's owner to think silently of any nonzero digit named.

Hand back the bill and tell the person to give the computer, in any order, all the digits except the one mentally selected. The computer screen remains blank. While the owner does this, you stand across the room with your back turned. Thus you have no inkling of what digit he or she thought of.

You now return to the computer and type: "What digit is _____ thinking of?" (including the person's name in the blank space). The chosen digit instantly appears on the screen.

The secret: As you call off the digits at the start of the trick, mentally add them in your head, "casting out nines" as you go along. This is easily done by adding the digits in any two-digit sum and remembering only the single-digit result. For example, if the first number is 9 and the second is 8, you say to yourself 9 plus 8 is 17, and 1 plus 7 is 8. You remember 8. This is known as the digital root of the sequence.

You must secretly tell the computer this number. One way to do it is to press the letter key directly below the digit on the keyboard's top row. As before, you can do this just after the space between the first two words of the question. The computer subtracts from this digital root the digital root of the set of seven numbers earlier given to it by the bill's owner. (The computer has of course determined the digital root of those seven numbers before you send it the digital root of all eight.) If the result is positive, that is the number chosen. If zero or negative, the chosen number is the result plus 9.

I hope these four tricks suggest the flavor and variety of amusing digital magic that computers can perform. There are endless possibilities involving cards, coins, dice, matches, and other common objects, as well as words and numbers. In many cases it is only a question of adapting a well-known magic trick to a computer presentation.

Chapter 10
Variations on the 12345679 Trick

An old number stunt that still never fails to entertain children starts with 12345679 (note the missing 8) in the readout of the calculator. Ask a child to tell you his favorite digit of 1 through 9; call this digit A. Tell him to multiply 12345679 by A, and then that product by 9. The result is AAAAAAAAA.

I recently amused myself by working out some variations on this chestnut. For example, the trick generalizes easily to chosen numbers of more than one digit, such as starting with 1122334455667789. Select any two-digit number, preferably with different digits; call this number AB (positional notation). Multiply the big number by AB, and then by 9, and you get an ABABAB... string.

The original stunt works, of course, because 9 times 12345679 equals 111111111. The generalization works because 9 times the starting number is 1010101.... Now for some variations based on numbers obtained by dividing repunits (1111111...) by an integer until there is no remainder.

Put 37037 in the readout. Multiply by A, then by 3, to get AAAAAA. The number you start with can be a long number that repeats 37037037....

Put 8547 in the readout. Multiply by A, then by 13, to get AAAAAA. Or use 8547008547008547....

Put 3003003 in the readout, or a number based on 3003003003.... Multiply by A, then by 37.

For 2849 (or 2849002849...), multiply by A, then 39.

For 271 (or 27100271...), multiply by A, then 41.

For 1221 (or 1221001221...), multiply by A, then 91.

Because the overwhelming number of repunits are composite, there is an infinite number of starting numbers that operate in this fashion.

This article first appeared in the *REC Newsletter* (Dec. 1991).

Here is a different variation, this one based on the prime factors of 12345679; 37 and 333667: Put 333667 in the readout. Multiply by A and then by 333 to get AAAAAAAAA. This works because 333 = 9 × 37 with 37 a factor of 12345679.

Of course, you can do this stunt in reverse: Put 37 in the readout. Multiply by A and then by 3003003 (the product of 333667 and 9) to get AAAAAAAAA.

There is also an infinity of starting numbers that can be used for generating an ABABAB... string. They are obtained by dividing the numbers 10101... by integers until there is no remainder. Examples:

Start with 1443 (or 1443001443...), multiply by AB, and then by 7. Or start with 3367 (or 3367003367), multiply by AB, and then by 3. Because it is easy enough to multiply AB by 3 mentally, you can request that 3367 be multiplied by the result (3 × AB). This generates the ABABAB string.

Start with 777 (or 777000777), multiply by AB, and then by 13.

A monstrous "magic" number lends itself to a demonstration of lightning calculation. Write down 9182736455463728191. (To aid memory, note the structure of complementary digits summing to 10 in pairs, followed by a final 1.)

Ask somebody to call out a two-digit number, AB, the two digits preferably different. In your head, multiply AB by 11 (easy enough) and request that the giant number be multiplied by this product (11 × AB). Of course, you'll need somebody to have computer access to verify the result of an ABABAB... string.

Chapter 11
More Calculator Whimsies

Pocket and desk calculators are not only useful, they also can be used to surprise and entertain yourself and your friends. Here is a choice selection of calculator amusements.

1. Select any number key (other than 0) and press it three times. Divide the number on display by 3, then divide the result by the number on the key you first punched. The result? 37.

2. Put on display any digit from 1 through 8. Divide by 9. The answer will "stutter" your original number. Clear the display, and enter a number between 10 and 98. Then divide by 99. The calculator will again repeat the selected number. Now clear and enter a three-digit number less that 999, and divide by 999. Guess what?

3. Punch in 987654312. Note that the 1 and 2 are in the wrong order. Divide by 8. The answer will surprise you! (If your calculator hold fewer than nine digits, you'll have to do this one by hand.)

4. Plug in a two-digit number, like 30. Reverse the digits (03, or just 3) and add the new number to the one you picked. You get a palindrome—a number that reads the same forward and backward. With some numbers, you need to reverse the sum and add again before you get results. For example, 39 + 93 = 132, and then 132 + 231 = 363. You may need to repeat these steps (reversing and adding) many times before you get a palindrome. Warning! *Don't even try 89 with a calculator.*

The palindrome of 89 is 8,813,200,023,188 (whew!). And just in case you're interested, 196 is the smallest number that can't be

This article first appeared in *Zigzag*, (Sept. 1995).

turned into a palindrome this way. Mathematicians have tried reversing and adding its digits more than 4,000 times without success!

5. You'll need a regular, six-sided die (like you can find in many games) for this trick. Give a friend the die and a calculator. Ask your friend to roll the die, while you have your back turned so you can't see it. Now tell your friend first to multiply the number on top of the die by 999999, and then to divide the result by 7.

   Even though you don't know the number on the die, you can call out, in random order, each digit in the product. The secret? Call out, in any sequence you like the digits 1,2,4,5,7, and 8.

6. Write "3990" on a piece of paper and place it facedown on a table. Ask one person to enter the year he or she was born on the calculator. Ask a second person to add his or her year of birth. Ask the first person to add the age he or she will be at the end of this year. Then have the second person do the same. Turn over the paper to show your prediction. It matches the total in the calculator!

7. Select any row of three number keys and press them in any order. To this number add a three-digit number obtained by punching, in any order, the keys in another row. Add a third number by punching in the keys in either diagonal in random order. Add another number using the keys in the other diagonal in random order. You should now have a four-digit number.

   Write this number down, then add all of its digits. If the sum is more than one digit, add the two digits to get a single digit. Amazingly, you will find that the digit on display is 6.

8. When certain numbers in the readout are viewed upside down, they make words. Here are three such tricks. In each case, after you do the math, turn your calculator around to read what it says.

   Say "hi" to the machine. Then divide 6.1872 by 8.

   What's the capital of Idaho? Multiply 8777 by 4.

   What did Santa Claus say when Rudolph showed him one of these stunts? Multiply 0.06734 by 6.

   What do you call people who can't multiply 6 by 3? Try 19,336 times 3.

On most calculators, the upside down readouts will look a lot like hELL.O, BOISE, and hOhOh.O. Some calculators, such as those built into a home computer, may not display the zero to the left of the decimal point in an answer, preventing the first and third of these tricks from working.

9. Enter 98765432. Divide by 8. The result will be 12345679, with the eight strangely missing. Multiply this number by your favorite digit between 1 and 9. Then multiply that result by 9. You'll get your favorite digit repeated over and over again.

Postscript

Here are some more computer tricks.

10. Put 1443 on display. Ask someone to tell you her age. (She must be older than 9.) Multiply 1443 by her age, then multiply by 7. The computer will "stutter" her age.

11. You'll need four paper matches for this trick. Think of a number from 1 through 9, then put it on display. Place a match on top of the *row* of keys that contain the number, and put another match on the *column* that contains the number.

To the number in the readout, add any digit represented by an *uncovered* key. Again, cover its row and column with two more matches. One key will remain uncovered. Add this number to the previous sum.

Even though the three digits were selected at random, the final sum will be 15.

12. Here's a simple game for two players. The first player pushes any key to enter a digit. The second player adds any digit represented by a key *adjacent* to the key previously pushed. (Adjacent means next to the key either to the left or right, or above or below, or diagonally.) Players take turns adding adjacent digits. The winner is the first to go above 50.

You can vary the game by choosing other numbers for the goal.

13. This is a way to prove you are psychic. Ask a friend to think of any three-digit number. Call it ABC. Tell him to put the number on display twice to make a number of the form ABCABC. (For example, if he chose 385, he would enter 385385.) While he does this, stand with your back turned so you can't see the number he selected.

"I'm starting to get some vibes from your number," you say. "They tell me it's exactly divisible by the unlucky number 13. Please divide by 13 and tell me if I'm right." You are. There is no remainder.

With your back still turned, say, "I have a strong impression that the number now on display is exactly divisible by the lucky number 11." When he divides by 11, sure enough, again there is no remainder.

Rub your forehead as if concentrating. Predict that if he divides by the lucky number 7, once more there will be no remainder. Right again!

Tell him to take a good look at the number in the readout. It will be ABC, the number he first thought of!

14. Ask someone to put on display any four-digit number. In your mind, subtract 2 from this number, then put 2 in front of the result. For example, assume his number is 3621. Subtracting 2 gives 3619, and putting 2 in front of it produces 23619. Write this number on a piece of paper and turn it face down.

Ask your friend to put a second four-digit number under the previous one. You then supply a third four-digit number, apparently a random one. Actually, for each digit you secretly select a digit that when added to the one directly above it will total 9.

Your friend now writes a fourth four-digit number. You write a fifth, again choosing digits the way you did before. The final addition problem will look something like this:

$$3621$$
$$9435$$
$$0564$$
$$7291$$
$$\underline{2708}$$

Total the five numbers, using the calculator. Turn over the paper to reveal your correct predication.

15. Write down any three-digit number provided its first and last digits are different. Reverse the digits to form another number. Take the smaller from the larger. If the result has two digits, put a zero in front. Reverse this number and add it to the previous one. Surprisingly, the answer will be 1089.

The number 1089 has many curious properties. For example, multiply it by any digit except 5. Put the reverse of this new number on display, divide by the number directly opposite the last key you pushed. For example, if you multiplied by 7, divide by 3. You are back to 1089!

16. While your back is turned, ask a friend to write down the number on any dollar bill, then scramble the digits any way she likes to make a second number. Using a calculator, she subtracts the smaller number from the larger.

With your back still turned, ask her to fix in her mind any digit not zero in the answer, then slowly call out every digit except the one she selected. As soon as she finishes, you name the omitted digit!

Secret: As she calls out digits, add them in your head. Whenever the sum goes above one digit, add the two digits and remember only the sum. For example, if she calls out 9 and 8, add them to get 17, then add 1 and 7 and remember 8. After she has called all the digits except the chosen one, subtract the single digit you have from 9. The result will be the digit she left out!

17. Enter 2143 in your calculator. (Note that this number is a permutation of 1234). Divide by 22, then hit the square root key twice. The readout will give π correct to eight decimal digits! The approximation is based on an identity discovered by Srinivana Ramanujan in 1914.

Chapter 12
Kasparov's Defeat by Deep Blue

Herbert Simon: "I would call what Deep blue does thinking."
John Searle: "Baloney."
   —quoted by Bruce Weber in "A Mean Chess-Playing Program Tears at the
Meaning of Thought," *New York Times* (February 19, 1996).

Imagine a person conversing with two hidden entities, one a human being, the other a computer. If the person is unable to decide which is which, we shall be forced to admit that the machine has achieved human intelligence. That was the so-called Turing test, proposed by British mathematician Alan Turing as long ago as 1950.

He predicted that by the year 2000, computers would speak fluently enough to deceive an "average interrogator" at least 30 percent of the time after about five minutes of dialogue. This cautious prophecy may well have become true. But will computers ever advance to a state at which their conversation, over a long period of time, will deceive even intelligent interrogators?

Today that question sharply divides mathematicians and artificial intelligence researchers. Many AIers believe that as computers grow in complexity and power it is only a matter of time until they become aware of their existence, with an intelligence that may even surpass ours. Mathematicians consider such predictions hogwash. To them the computer is no more than a tool for juggling numbers so rapidly that it is no longer necessary to waste hours making large calculations by hand; it is only speed, accuracy and flexibility that distinguish a brainless computer from a brainless abacus.

On Sunday, May 11, IBM's chess playing computer Deep Blue won a six-game match in Manhattan against Russia's Gary Kasparov, the world champion believed to be the best chess player ever. Kasparov won one game, tied three, and lost two. The final game was a crushing

---

This essay first appeared in *The Washington Post* (May 25, 1997).

defeat. It lasted about an hour and ended after nineteen Blue moves. Kasparov's blunder was accepting Blue's sacrifice of a knight for a pawn. He failed to see what Deep Blue had seen, that by losing a knight, Blue obtained an overwhelming positional advantage.

Kasparov was a shaken and angry loser. After apologizing for his poor play, he immediately launched into a bitter attack on Deep Blue, challenging it to enter tournaments and "play real chess." He "guaranteed" that if this happened he would "tear Deep Blue to pieces."

Kasparov's defeat was hailed as a milestone, not only in chess history but also in the progress of AI. In my opinion it was a very minor milestone.

Exactly what do computers do? They are mindless machines designed to manipulate binary digits—ones and zeros—modeled by electrical impulses switched here and there along wires. The simplest example of such a device is the abacus. Ones are modeled by beads, zeros by empty spaces along rods. Switches are provided by fingers that slide the beads according to algorithms—procedures that give instructions to the fingers. Muscles of the hand and arm furnish the energy. Of course, the power of an abacus is severely limited by the small number of rods and beads and by the long time it takes to operate the device.

Mechanical computers are more efficient, They can be made with cogwheels, with jets of water flowing through a network of tubes, with levers and pulleys, with balls rolling down inclines—indeed with almost anything that can be manipulated by energy. I have a cardboard device, printed years ago as an advertising premium, that plays unbeatable tic-tac-toe by using a sliding strip and a rotating disk. Not long ago a group of clever computer hackers built a tic-tac-toe machine with tinker toys. In principle, a tinker-toy machine can do everything a supercomputer can do—provided it is large enough and given enough time.

Supercomputers differ from mechanical calculators in only one fundamental way: By using electricity and tiny silicon switches to move ones and zeros through wires networks it gains incredible speed. If you call what it does "thinking," you might just as well say the beads of an abacus are thinking while they add numbers.

A supercomputer's awesome speed enables it to answer mathematical questions no one could answer by hand. No human mind could have calculated, as computers have easily done, π to millions of decimal digits. The fact that computer programs can play grandmaster chess is no more surprising than their ability to multiply gigantic numbers faster than any human lightning calculator. Deep Blue defeated Kasparov in a totally mindless way. It no more knew it was playing chess than a vacuum cleaner knows it is cleaning a rug. It cares not a whit whether it wins or loses.

Human chess players examine a few future moves at a rate of several per second, using experience and intuition to avoid considering irrelevant moves. Deep Blue examines *all* possible future positions for 10 or more moves ahead at a rate of 200 million positions a second. It is this fantastic speed, combined with "selectivity rules" for rating positions, that gives Deep Blue its enormous brute-force power. And it wins games. As is often pointed out, airplanes fly faster than birds but without flapping their wings.

For years, chess programs have defeated grandmasters when moves must be made rapidly. And the present checker champion of the world is a computer program. Checkers is so much simpler than chess that in a decade or two the game may be solved—a program will play a perfect game.

Chess is far from being solved. But computers have passed a Turing chess test. A grandmaster cannot know whether his hidden opponent is another grandmaster or a computer program. But that achievement is a far cry from the complexity of human intelligence.

"Complexity" has become a buzzword, precisely defined in computer science. Philosophers have broadened the term to apply to the evolution of the universe after it exploded into existence. Although the universe as a whole is increasing in entropy (disorder), there are regions here and there where disorder gives way to beautiful order in the emergence of ever more complex systems. The formation of galaxies, stars, and planets are striking examples. On at least one planet, life has emerged and evolved in the direction of ever-increasing complexity, culminating in the brains of such strange creatures as you and me.

Now the notion that, as complexity increases, astonishing new properties emerge is as old as the ancient Greek thinkers. Wondrous properties appeared when atoms formed from quarks and electrons. Even more amazing properties emerged when atoms joined to make molecules. Hydrogen and oxygen are simple elements with simple properties. Put them together and you get water, a substance with remarkable attributes unlike those of either element, properties that may be absolutely essential for the emergence of life.

So will computers, of the sort we know how to build, ever rival the complexity of human intelligence? In 1988, Hans Moravec, who heads a robotics laboratory at Carnegie Mellon University, wrote a book titled *Mind Children: The Future of Robot and Human Intelligence* in which he predicted that computers would be surpassing human minds in less than half a century. Similar views, though with much longer time frames, have been advanced by AI researchers Marvin Minsky, Herbert Simon, and Douglas Hofstadter. Philosopher Daniel Dennett is not in the least mystified by consciousness, believing he explained it—and that computers will one day have it—in his best-selling 1991 book *Consciousness Explained.*

Moravec, physicist Frank Tipler and a few others actually believe that computers will eventually render the human race obsolete. They will become our "mind chidren," destined to take over the tak of colonizing the cosmos as humanity goes the way of the dinosaurs.

Moravec is the strongest of what are called "strong AIers." He writes, "Today our machines are still simple creatures.... But within the century they will mature into entities as complex as ourselves, and eventually into something transcending everything we know."

The words have a familiar ring. Here is Samuel Butler writing in his 1872 fantasy "Erewhon": "There is no security against the ultimate development of mechanical consciousness. Reflect upon the extraordinary advance which machines have made during the last few hundred years, and note how slowly the animal and vegetable kingdoms are advancing.... The present machines are to the future as the early Saurians to man."

The difference between Butler's remarks and similar sentiments by hard AIers in that Butler was writing satire.

Opposing these wile fantasies is a group of thinkers sometimes called "mysterians," and rightly so, because they believe that our minds remain a profound mystery. Here the mysterians, among whom I count myself, join the mathematicians who work with computers. Among the most outspoken mysterians are three American philosophers, John Searle, Thomas Nagel, and Colin McGinn, the British mathematical physicist Sir Roger Penrose, and psychologist Steven Pinker. Penrose's two books, *The Emperor's New Mind* and its sequel *Shadows of the Mind*, are the strongest attacks yet on the belief that computers will soon cross a threshold of complexity making them aware of who they are, able to feel pleasure and pain, to create and laugh at jokes, to love and hate, to make moral decisions, to write great poetry, music and novels, to make new scientific discoveries, and to meditate on philosophical and theological questions.

Most mysterians do not believe a "soul" exists apart from the brain. They accept the view that our "self," with its consciousness and free will (two names for the same thing), is a function of a material brain, a computer made of meat, as Minsky likes to say. They contend that our brains are so much more complicated than today's computers that we have only the dimmest comprehension of how they operate.

Neuroscientists are making progress, but as yet they do not even know how memories are stored and retrieved. Penrose simply insists that the abilities of the human mind are not going to emerge from computer complexity as long as computers consist of nothing more than electric currents moving through wires in a manner dictated by software. Until we know more about how brains do what they do, we will not be able to construct computers that will come close to rivaling human minds. Penrose even contends that no such computers will be built until we know more about laws of physics deeper than quantum mechanics.

Deep Blue's defeat of Kasparov in no way signals the emergence among computers of anything faintly resembling human intelligence. What Deep Blue does is nothing qualitatively different from what an old-fashioned adding machine does. It merely twiddles numbers faster than a mechanical machine. Perhaps some day, if quantum computers are ever made operational, they will be on their way toward something resembling human thought.

I once wrote that Moravec suffers from having read too much science fiction. A prominent science fiction author took me to task for this remark, insisting that no one can read too much science fiction. What I meant, of course, was that Moravec took too uncritically the stories he read about the coming of intelligent robots. It is a long, long distance from the circuitry of Deep Blue to the mind of a mouse.

Chapter 13
Computers Near the Threshold?

The notion that it is possible to construct intelligent machines out of nonorganic material is as old as Greek mythology. Vulcan, the lame god of fire, fabricated young women out of gold to assist him in his labors. He also made the bronze giant Talus, who guarded the island of Crete by running around it three times a day and heaving huge rocks at enemy ships. A single vein of ichor (the blood of the gods) ran from Talus's neck to his heels. He bled to death when he was wounded in the ankle or, according to another myth, when a brass pin in his heel was removed.

After the Industrial Revolution, with its wonderful machinery, writers began to speculate about the possibility that humans as well as gods could some day build intelligent machines. In Chapters 23, 24, and 25 of his novel *Erewhon* (1872), Samuel Butler wrote about the coming of such robots. Tik-Tok, one of the earliest mechanical men in fiction, was a windup copper person who made his first appearance in L. Frank Baum's *Ozma of Oz* (1907). He was manufactured in Ev, a land adjacent to Oz, by the firm of Smith and Tinker. A plate on his back said that the robot "Thinks, Speaks, Acts, and Does Everything But Live."

After the computer revolution produced electronic calculating machines, with their curious resemblance to the electrical networks of human brain, the possibility of constructing intelligent robots began to be taken seriously, especially by leaders of AI (artificial intelligence) research, and by a few fellow traveling philosophers. Hans Moravec directs a robot laboratory at Carnigie Mellon University; in his book *Mind Children* (1988), he predicted the appearance of robots with human intelligence before the end of the next fifty years. Both he and Frank Tipler, a Tulane University physicist, are convinced that com-

This essay first appeared in *Mysteries of Life and The Universe.* (1992), edited by William H. Shore.

puters will soon *exceed* human intelligence, making the human race superfluous. Computers will then take over the burden and adventure of colonizing the universe.

Here is a passage from *Erewhon* that could have been written seriously by Tipler or Moravec:

> There is no security against the ultimate development of mechanical consciousness, in the fact of machines possessing little consciousness now. A mollusc has not much consciousness. Reflect upon the extraordinary advance which machines have made during the last few hundred years, and note how slowly the animal and vegetable kingdoms are advancing. The more highly organized machines are creatures not so much of yesterday, as of the last five minutes, so to speak, in comparison with past time. Assume for the sake of argument that conscious beings have existed for some twenty million years: See what strides machines have made in the last thousand! May not the world last twenty million years longer? If so, what will they not in the end become?

Another passage, remarkably prophetic, from the same book:

> Do not let me be misunderstood as living in fear of any actually existing machine; there is probably no known machine which is more than a prototype of future mechanical life. The present machines are to the future as the early Saurians to man. The largest of them will probably greatly diminish in size.

Actually, these are not the narrator's words but sentences that Butler attributes, tongue firmly in cheek, to an Erewhonian professor. The professor's opinions prompt the Erewhonians to destroy all their machines before they surpass human intelligence and take over the world.

In *Mind Children* Moravec puts it this way:

> Today our machines are still simple creations, requiring the parental care and hovering attention of any newborn, hardly worthy of the word *intelligent*. But within the next

century they will mature into entities as complex as ourselves, and eventually into something transcending everything we know—in whom we can take pride when they refer to themselves as our descendants.

The most powerful attack on such opinions, which have come to be called "strong AI," is Roger Penrose's best-seller *The Emperor's Mind* (1989). Naturally, the book was vigorously lambasted by strong AIers. Because I wrote the book's foreword, I too have been denounced for my obtuseness. This essay is an effort to set down in more detail precisely what I believe about the possibility that computers will soon be able to converse with us in ways indistinguishable from the conversations of human beings.

First, I should make clear that I am not a vitalist who thinks there is a "ghost in the machine"—a soul distinct from the brain. I believe that the human mind, like the mind of any lower animal, is a function of a material lump of organic matter. Although I remain open to the Platonic possibility of a disembodied soul, as I am open to any metaphysical notion not logically contradictory, the evidence against it seems overwhelming. Strong arguments for a functional view of the mind are too familiar to need summarizing.

If a human has a nonmaterial soul, it is hard to see why the same should not be said of an amoeba, a plant, or even a pebble. A few pan-psychic monists such as Charles Hartshorne actually do say this, but I consider it an absurd misuse of words, a "category mistake," to talk of a potato in a dark cellar as having what Butler called "a certain degree of cunning." Here are two pan-psychic quotations from Butler's imaginary Erewhonian professor:

> But who can say that the vapor engine has not a kind of consciousness? Where does consciousness begin, and where end? Who can draw the line? Who can draw any line? Is not everything interwoven with everything? Is not machinery linked with animal life in an infinite variety of ways?
>
> Shall we say that the plant does not know what it is doing merely because it has no eyes, or ears, or brains? If we say that it acts mechanically, and mechanically only, shall we not be forced to admit that sundry other and apparently very deliberate actions are also mechanical? If it seem to us that the plant kills and eats a fly mechanically, may it not seems to the plant that a man must kill and eat sheep mechanically?

I agree with Aristotle that the self is the "form" of the body, or in modern terminology, a pattern of the molecular structure of organic matter inside our skull. Of course the pattern is far more complex than the pattern of a vase or the Empire State Building.

Life did indeed evolve along continua, but there are spots (albeit with fuzzy edges) where wide chasms were crossed and new properties of matter emerged. The first great threshold was the emergence of life from lifeless compounds. And the last of the great thresholds, the greatest of them all, was the evolution of the brain with such properties as consciousness (self-awareness), expanded free will (with all its moral implications), a sense of right and wrong, a sense of humor, the power to communicate complicated ideas by speech and writing, and a raft of creative skills such as the abilities to compose poetry and music, paint pictures, discover significant mathematical theorems, and invent scientific theories capable of being tested. With the last skill came awesome power to control the process of evolution and steer it in new directions, as well as the power to terminate the process.

It goes without saying that many of these human traits are possessed to a weak degree by animals. Chimpanzees seem to have a low-level awareness of themselves and an ability to make decisions. (*Free will* and *self-awareness*, by the way, are for me two names for the same phenomenon, like Einstein's principle of the equivalence of gravity and inertia, or, what amounts to the same thing, the equivalence of gravitational and inertial mass. I cannot imagine myself having free will without being self-aware, nor can I conceive of being self-aware without some degree of free will). Monkeys have a feeble sense of humor; you can watch them play pranks on each other in a cage. Bower birds may have a dim sense of visual beauty. Apes can communicate with humans by signs, and so can a dog or cat. A chimp can make and test conjectures about how to get a banana from the ceiling if there are boxes lying around. And so on. That animals feel emotions of love and pain is undeniable. It is equally undeniable that a major gulf of some sort was crossed when humans evolved from bestial ancestors.

The question is not whether our human traits emerged as a function of an evolving brain, as I assume they did, but whether it will be easy or difficult, perhaps even impossible, to build a calculating machine complex enough to leap the same threshold. At this point we touch the central theme of Roger Penrose's brilliant, controversial book *The Emperor's New Mind*.

The difficulty in crossing the threshold, Penrose argues, is that we don't yet know enough about matter to know how to do it. Clearly we know far more about particles that did Democritus, but we are still a long way from understanding those particles. In standard theories, matter is made of leptons and quarks, and these particles are taken to be geometrical points, or at least *pointlike*, as physicists prefer to say. In recent superstring theories they are not points but inconceivably tiny loops. In either case, Newton's hard little pebbles that bounce against one another—the kind of matter Bishop Berkeley ridiculed as a "stupid, thoughtless somewhat"—have now totally dissolved. What is left is mathematics. Leptons and quarks, whether points or loops, are not made of anything. Their fields are just as ghostly. On the quantum level, to put it bluntly, there is nothing except mathematical patterns.

My readers know how impatient I am with some pragmatists, phenomenologists, and subjective idealists of various schools who heap scorn on the notion that mathematical structures are "out there" with a reality that is not mind-dependent. For these thinkers, mathematical reality is located within human experience. Like Penrose and the overwhelming majority of eminent mathematicians past and present, I am a Platonist in the sense that I believe mathematical patterns are discovered, not invented. Of course, they are still invented, in a sense. Everything humans do and say is what humans do and say. Mathematics obviously is part of human culture, but to say so is to say something utterly trivial. The fact that only humans can talk or write about mathematics and laws of physics does not mean that it is useful to deny that mathematics and laws of physics are embedded in an enormous universe not made by us, but of which we are a part, and an inconceivably tiny part at that.

As I have said before, if two dinosaurs met two others in a forest clearing, there would have been four dinosaurs there—even though the beasts were too stupid to count and there were no humans around to watch. I believe that a large integer is prime before mathematicians prove it prime. I believe that the Andromeda galaxy had a spiral structure before humans arose on Earth to call it spiral. As the noted Bell Labs mathematician Ronald Graham recently put it, mathematics is not only real, it is the *only* reality.

Some eccentric philosophers prefer to think that human minds alone are really real. There are even physicists, overwhelmed by the solipsistic tinges of QM (quantum mechanics), who like to talk the same way. But the human mind is made of molecules, which are in turn made of atoms, which are in turn made of electrons, protons, and neutrons. The protons and neutrons are made of quarks. What are quarks and electrons made of? Nothing except equations. Let's face it. You and I, at the lowest known level of our material bodies, are made of mathematics, pure mathematics, mathematics uncontaminated by anything else.

The most elegant theory of matter today is of course QM. Unfortunately, it is riddled with mysterious paradoxes. In recent years Einstein's EPR paradox (named with the initials of Einstein and two of his colleagues) has been the most debated. How can the measurement of one particle cause the emergence of a property on a correlated particle that can be millions of light-years away?

It seems to happen either instantaneously or with a speed faster than light can travel between the particles. In the first case, the paradox violates the dogma that prohibits instant action at a distance. In the second case, it seems to violate relativity, which prohibits information from traveling faster than light.

None of the many proposed resolutions is satisfactory. The many-worlds interpretation of QM seems to get rid of the paradox, but there is an enormous price to be paid. One must posit billions upon billions of ever-proliferating parallel universes in which everything that can happen, does. Other efforts to solve the EPR paradox do no more than restate it in a different language. It is no good, for example, to say that the two correlated particles are part of a single quantum system whose wave function (or state vector in another language) collapses all at once, so naturally when you collapse it by measuring one particle you obtain information about the other. This simply restates the formalism. You now have to explain how two particles, light-years apart, can remain correlated.

For years Penrose has maintained, along with David Bohm, Paul Dirac, Erwin Schrödinger, and other great physicists, that QM is not the ultimate theory of fields and particles. This was Einstein's own view. Indeed, it was Einstein who first proposed the notorious EPR paradox in an effort to show that QM was incomplete. Working physicists, for the most part, never worry about such things. As long as QM works, and of course it works magnificently, they simply accept the fact that (as Richard Feynman liked to say) QM is "crazy." Don't try to understand how it works, Feynman warned his students, because nobody knows how it works. Should physicists leave it at that? No, insist Bohm and Penrose, because that tends to discourage research that may some day find that QM, like Newton's gravity, is only a good approximation of a deeper theory. Penrose himself is trying to go deeper, with a geometrical theory of particles and fields about which I am not competent to have an opinion.

Penrose contends, and I agree, that until we know more about matter on a level beyond QM, we will not understand how our minds can be a function of our gray matter. Until we know those deeper laws, we will not even come close to constructing a machine that can do everything our minds can do.

In Penrose's opinion, the great mistake behind the optimistic predictions of strong AIers is the assumption that machines made of wires and switches, operating with algorithmic software, can cross the great threshold. Let's look at this assumption more closely. We know from the work of Alan Turing and others that it is possible in principle to build computers out of any kind of equipment that transmits energy along channels, with switches to tell the energy where to go. You can build computers with networks of pipes that hold a flowing liquid. You can build them with rotating gears, with string and pulleys, with little balls that roll down inclines or slide along wires as on an abacus. Mechanical devices of these sorts have been constructed in the past. If you are interested, you can read about them in my *Logic Machines and Diagrams*.

Every machine, the philosopher-mathematician Charles Peirce once observed, is a logic machine in having aspects that model logic functions. The blades of an eggbeater rotate in one direction "if and only if" you turn the handle clockwise. They go the other way if and only if you turn the handle counterclockwise. An old mechanical typewriter is a jungle of binary logic relations. (Peirce, incidentally, was the first to show how a simple logic machine handling binary functions could be built with electrical currents and switches.) A few years ago a group of computer hackers constructed a machine out of Tinker Toys that played perfect tic-tac-toe. There is no reason why, in theory, one could not build a Tinker Toy computer that could do everything a Cray computer can do or, indeed, what any supercalculating machine of the future could do. Of course, it would have to be monstrously large and intolerably slow. Would its sluggishness dilute its consciousness? Could it still write a great novel, provided it had a few thousand years to do it?

Now, no one in his right mind would say that a Tinker Toy tic-tac-toe machine "knows" it is playing tic-tac-toe any more than a vacuum cleaner knows it is cleaning a rug or a lawn mower knows it is cutting grass. Sophisticated computer programs that now play Master chess differ from tic-tac-toe programs only in the complexity of their algorithms. A computer with such a program is no more aware it is playing chess than an eggbeater is aware it is beating eggs.

Strong AIers believe that as computers of the sort we presently know how to construct keep growing in the complexity of their circuitry and software, they will eventually cross a threshold and become conscious of what they are doing. If one believes this, is not once forced to say that a Tinker Toy machine of comparable complexity, or even one made with rolling marbles, will cross the same threshold?

I admit that all this may someday be possible, but I agree with Penrose and such opponents of strong AI as the philosopher John Searle that it seems extremely unlikely. We know very little about how the brain of a fish or a bird works. We do not even know how memories are stored in the mind of an ant. It is true that electrical pulses are silently shifted about inside the skulls of animals, but this is done in a manner far from understood. What Penrose is telling us is that evolution, working on computers made of meat, crossed a threshold in a way that involves laws of physics not yet known. It could be that if and when those deeper laws are discovered and we know exactly how our brain does what it does, we will be able to construct a replica (perhaps made of nonorganic matter, perhaps requiring organic molecules) that will simulate a human mind. But to expect a calculating machine made with components of the sort now in use or imagined to cross the threshold seems hopelessly unwarranted.

What does a computer do? It twiddles symbols—symbols that are meaningless until we attach meanings to them. It twiddles them in blind obedience to syntactical rules provided by the software. But our minds do more than twiddle symbols. They also twiddle meanings of symbols. I can easily imagine a monstrous machine made of Tinker Toys that can play Grand Master chess, but I cannot imagine it will know it is playing chess. By the end of this century I expect a chess program to be able to defeat any grand master while playing under the usual time restraints. Even now chess programs can crush grand masters when moves must be made within a few seconds. I expect that powerful computers will steadily improve in their ability to do all sorts of extraordinary things, but these things will all be done by symbol twiddling. I do not believe that the complexity of their circuitry will push them across the magic threshold.

Endless novels, stories, plays, and even operas have been written about intelligent robots, and it is no accident, I suspect, that so many strong AIers were science-fiction buffs in their youth. My favorite novel about robotics is a little-known one by Lord Dunsany, inexplicably never published in the United States. Titled *The Last Revolution* (1951), it concerns a rebellion against humanity of super-intelligent, self-reproducing machines, the sort of rebellion that Tipler and Moravec believe possible. The book's funniest scene occurs when the narrator plays chess with the first prototype. Its inventor, Ablard Pender, lives with his aunt Mary. He pretends to wind up his "gadget" with a key so as not to frighten her by letting her know it is alive.

It requires only a few moves for the narrator, whose Ruy López opening quickly takes a bizarre turn not in any chess manual, to realize he is playing not only against an intelligence superior to his own but against a mind aware of what it is doing.

When Pender's girlfriend, Alicia, first sees the crablike, four-legged monster and its eyes like a cockroach's, there is an intuitive flash on her face like forked lighting. She senses immediately that the thing is alive. A dog, frightened by something it too knows is living but that has no smell, howls and bites the iron. The thing tears the dog to pieces.

The brain of Pender's robot consists of fine wire that transmits electrical pulses. "Did you make it entirely yourself?" Alicia asks.

"Yes, of course," Pender replies. "Don't you like it?"

"Time," says Alicia, "will have to show that."

Postscript

As this essay makes clear, I am among the ranks of those who are called "mysterians." We do not deny that the mind emerges from a complex pattern of molecules, but we believe that the brain's complexity is so vast that at present we simply do not understand how it produces self-awareness and free will. We also believe that computers, operating with wires and switches that shift electrical currents around a network in obedience to software algorithms, will never cross the threshold at which anything resembling the human brain will emerge.

Since writing this essay I came across an interesting 1872 letter from Samuel Butler to Charles Darwin, reprinted in Darwin's autobiography. In it Butler explains that the section in *Erewhon*, where he suggests that machines will soon make humans obsolete, was written in jest. Here is what Butler has to say:

> I venture upon the liberty of writing to you about a portion of the little book *Erewhon* which I have lately published and which I am afraid has been a good deal misunderstood. I refer to the chapter on Machines in which I have developed and worked out the obviously absurd theory that they are about to supplant the human race and be developed into a higher kind of life.
>
> When I first got hold of the idea I developed it for mere fun, and because it amused me and I thought it would amuse others, but without a particle of serious meaning; but I developed it and introduced (it) into *Erewhon* with the intention of implying "See how easy it is to be plausible, and what absurd propositions can be defended by a little ingenuity and distortion and departure from strictly scientific methods," and I had Butler's *Analogy* in my head as the book at which it should be aimed, but preferred to conceal my aim for many reasons. Firstly the book was already as heavily weighted with heterodoxy as it would bear, and I dare not give another half ounce lest it should break the camel's back; secondly it would have interfered with the plausibility of the argument, and I looked to this plausibility as a valuable aid to the general acceptance of the book; thirdly it is more amusing without any sort of explanation, and I thought the drier part that had gone before wanted a little relieving; also the more enigmatic a thing of this sort is, the more people think for themselves about it, on the principle that advertisers ask "Where is Eliza?" and "Who's Griffiths?" I therefore thought it unnecessary to give any disclaimer of an intention of being disrespectful to the *Origin of Species* a book for which I can never be sufficiently grateful, though I am well aware how utterly incapable I am of forming any opinion on a scientific subject which is worth a moment's consideration.

In L. Frank Baum's *The Scarecrow of Oz*, there is a scene in which a grasshopper rests on the Scarecrow's severed head. "Are you alive?" the grasshopper asks.

"That is a question I have never been able to decide," said the Scarecrow's head. "When my body is properly stuffed I have animation and can move around as well as any live person. The brains in the head you are now occupying as a throne, are of very superior quality and do a lot of very clever thinking. But whether that is being alive, or not, I cannot prove to you; for one who lives is liable to death, while I am only liable to destruction."

Chapter 14
Cornering the King

In a 1992 issue of Michael Ecker's *REC Newsletter* I posed the following chess problem. The queen is at the lower right corner of a rectangular field, and the king is at the upper left corner. On square rectangles the king is in check, so he moves first. Otherwise the queen goes first. Players alternate moves. On what fields can the queen force the king into the upper left corner? The problem generalizes a puzzle by Marek Penszko, of Poland, which appeared in *Games* magazine, the date of which I have lost.

The king clearly can't be cornered on $2 \times n$ fields or on a 3×3 square. He *can* be cornered on certain $3 \times n$ fields when $n > 3$. The smallest board of interest, therefore, is the 3×4 board.

Assuming the king always makes his best moves, what's the minimum number of queen moves required to corner the king on the 3×4? It's not as easy to solve as it seems. And what's the story on larger boards?

I conjectured in *REC Newsletter* that the king could not be cornered on any square board, though I had no proof. I also conjectured that the queen could win on all rectangular $n \times m$ boards, $n > 2$ and $m > n+1$.

Andy Liu, the ace problem creator and solver at the University of Alberta, Canada, was intrigued by the problem. He and I shared a byline on the following article, written by Liu:

A Royal Problem: And Alice Is Caught in the Middle

The Red Queen was furious, as usual. Her current ire was brought on by the absence of the Red King from his Palace. On her rare visits, she expected to see whom she had come to see.

This article first appeared in *Quantum* (July/August 1993).

"Bring the old fool back here, or else!" roared the Red Queen, who was related to the Queen of Hearts.

"Or else what?" asked Alice, but only after Her Majesty had swept radiantly out of earshot back to her side of the Palace.

"Off with your head!" Tweedledum said.

"What else?" added Tweedledee rhetorically.

"Oh, dear," said Alice, "this puts a new meaning to ten percent off the top. What shall I do? I don't even know where the Red King is."

The twins brought out a map of the land. It was the familiar 8 × 8 chessboard in Figure 1.

"I bet I know where His Majesty is," said Tweedledum.

"On h6!" exclaimed Tweedledee.

"How do you know that?" Alice asked.

"Well," said Tweedledum, "the Red King plays it safe. He never ventures out of his Kingdom into the Borderland."

"He also refuses to cross over to the Queen Side," added Tweedledee.

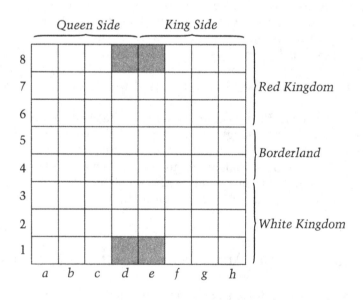

 Palace

Figure 1.

"So he is confined to twelve squares. That is helpful, but I still don't see how you can be so sure that he is on h6."

"His Majesty likes to be as far away from the Red Queen as he possibly can," Tweedledum said.

"Actually, as far from the Red Queen's Palace as possible," corrected Tweedledee. "He has no control over the whereabouts of Her Majesty."

"There is another problem," said Alice. "If the Red King does not want to come back to e8, how can I persuade him against his wishes?"

The twins thought for a while, and fought for a while just to pass the time. Then they both came up with a brilliant idea. Not surprisingly, it was the same idea.

"Are you in mortal fear of the Red Queen?" Tweedledum asked Alice.

"Of course. Who isn't?"

"Of all people, who fears her the most?" asked Tweedledee.

"Hard to say," Alice replied. Then it occurred to her. "The Red King, of course."

"Right!" said Tweedledum. "He could not risk getting caught in a mating situation with the White Queen."

"So if you disguise yourself as that good lady, you can drive His Majesty back here," declared Tweedledee triumphantly.

"It is worth a try," said Alice, somewhat encouraged. "I should not waste any time by venturing outside of those twelve squares either."

"Make sure you don't corner His Majesty on h8," Tweedledum advised Alice.

"Also, do not drive him into the Borderland," said Tweedledee. "His Majesty may find out that it is not as dangerous as he makes it out to be."

"Well, I'd better hurry and bring His Majesty back as soon as I can. The Red Queen's patience is shorter than her temper!"

Problem 1.

On the miniature chessboard in Figure 2, White has a lone Queen on e8, and Red has a lone King on h6. White moves first, and wins if the Red King is driven back to e8 within 10 moves. If this is not accomplished, then Red wins. Other than what is noted above, normal chess rules apply. With perfect play, which royalty wins?

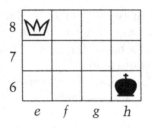

Figure 2.

Alice was able to accomplish her mission, only to have the Red King slip out again. Humpty Dumpty, in his lofty position on the wall, spotted His Majesty on h4 this time.

Alice correctly deduced that the Red King still harbored no thought of crossing over to the Queen Side. While he had temporarily conquered his fear of the Borderland, he was not yet willing to venture into the White Kingdom.

Having lost much time in accomplishing her first mission, Alice set out immediately to reenact the drama, but on an enlarged stage.

Problem 2.
On the miniature chessboard in Figure 3, White has a lone Queen on e8 and Red has a lone King on h4. White moves first, and wins if the Red King is driven back to e8 within 14 moves. If this is not accomplished, then Red wins. Other than what is noted above, normal chess rules apply. With perfect play, which royalty wins?

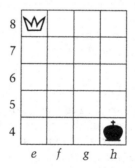

Figure 3.

Alice drove the Red King back to his Palace just in time.

"Come along," roared the Red Queen. "We have to attend a summit conference with the White Queen and her consort."

"What is the matter this time, dear?" asked the Red King timidly.

"We have been discussing the partition of the Borderland. There is too much goings-on here, especially on h4, or so I hear."

"I can't imagine what," murmured the Red King.

"Anyway, the White Queen and I have agreed to establish our borders between ranks 4 and 5. We just meet to formalize the deal."

"If you say so, dear."

As soon as the new treaty was signed, the Red King headed for h5, the furthest haven within his domain. Alice was dispatched after him a third time.

Problem 3.

On the miniature chessboard in Figure 4, White has a lone Queen on e8, and Red has a lone King on h5. Red moves first because the King is already in check. White wins if the Red King is driven back to e8 within 12 moves. If this is not accomplished, then Red wins. Other than what is noted above, normal chess rules apply. With perfect play, which royalty wins?

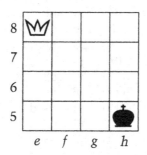

Figure 4.

Solutions

1. White Queen wins as follows:

(1)	f8	h7 (a)
(2)	f6	g8
(3)	h6	f7
(4)	h7	f8 (b)
(5)	g6	e7
(6)	g7	e6
(7)	h7	f6
(8)	g8	e7
(9)	g6	f8
(10)	h7	e8

Notes: (a) If (1) ... g6, then (2) h8 f7, and continue from (4).
  (b) If (4) ... e6, continue as before. If (4) ... f6, continue from (8).

2. White wins as follows:

(1)	e5	g4
(2)	f6	h5
(3)	f4	g6
(4)	e5	h6 (a)
(5)	f5	g7
(6)	e6	h8 (b)
(7)	h6	g8
(8)	g6	h8 (c)
(9)	g5	h7
(10)	f6	g8
(11)	h6	f7
(12)	g5	f8 (d)
(13)	g6	e7
(14)	f5	e8

Notes: (a) If (4) ... f7, continue from (12). If (4) ... h7, continue from 10.
  (b) If (6) ... f8, continue from (13). If (6) ... h7, continue from (10).
  (c) If (8) ... f8, then (9) h7 e8.
  (d) If (12) ... e6, then (13) f4 e7, and continue from (14).

3. Red King wins. An article in the next issue will present an argument supporting this conclusion by solving the Royal Problem for all $m \times n$ chessboards, $m \geq n \geq 3$.

Postscript

The general problem was completely solved by Andy Liu's amazing students. Three of them, Jesse Chan, Peter Laffin, and Da Li, all in tenth grade, gave their solution in "Martin Gardner's 'Royal Problem,'" in *Quantum* (September/October 1993). They confirmed my conjecture that the king can't be cornered on square fields, but shot down my other conjecture, that the queen wins only on rectangular boards when $n > 2$ and $m > n + 1$. I thought I had shown that the king wins on the 4×5 board, but Andy's students surprised me by finding a queen win on this board.

Perhaps I should add that all chess rules apply. If the queen is adjacent to the king, the king can take the queen, thereby winning the game. And the king wins if he is stale-mated on a square outside the upper left corner, or if he achieves perpetual check outside that corner.

Chapter 15
Toroidal Currency

I doubt if many mathematicians outside Australia are aware that the Reserve Bank of that nation has issued two bank notes that are toroidal. They are the $5 bill (Figure 1) printed in 1992 and the $10 bill (Figure 2) printed in 1993.

The front and back of each of these handsome notes are shown. Observe that on both sides of each bill the patterns at top and bottom "wrap around," as well as the patterns on the left and right edges. As all topologists know, if the right and left sides of a rectangle, and its top and bottom, are joined, the result is the familiar torus, or doughnut shape. If just one pair of sides is connected by reversing one of the edges, the structure is a Klein bottle. If the reversal applies to both pairs of edges, the structure is a projective plane. Perhaps some day a nation will print projective plane or Klein bottle notes!

The purpose of the toroidal wraparounds is to make it harder to counterfeit the bills. Sophisticated color copiers have made counterfeiting much easier all over the globe.

An additional anticounterfeiting feature of each note is a transparent "window" at a lower corner (it appears black as printed here). Still another such device is the little circle showing four points of a star on one side and three points on the other. Hold either bill up to a strong light and the points fuse to form a perfect seven-pointed star, symbolizing Australia's seven original states. The slightest variation in register would distort the star.

The face on the $5 note obviously is that of Queen Elizabeth. The woman on the $10 note is Dame Mary Gilmore, an Australian poet who worked tirelessly to battle injustices in the nation, especially in the treatment of the native aborigines. The man on the opposite side is A. B. "Banjo" Paterson, a ballad singer and journalist best known for having written the words of "Waltzing Matilda," Australia's un-

---

This article first appeared in *Quantum* (September/October 1994).

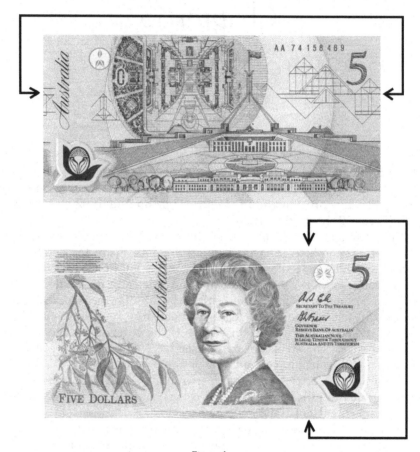

Figure 1.

official national anthem. (Matilda, by the way, is not a woman but a knapsack.) "The Man from Snowy River" is another of Paterson's popular songs.

The $10 note has been made 7 mm longer than the $5 bill to help sight-impaired persons distinguish between the two. Both notes are made of locally produced polymer rather than imported paper. The polymer lasts longer, stays cleaner, and can be recycled for plastic products.

Melbourne artist Max Robinson designed the new $10 note. Behind Paterson's profile, in microprinting, are lines from Paterson's verse, making the bill even more difficult to counterfeit.

Figure 2.

Chapter 16
Six Challenging Dissection Tasks

Karl Scherer, a computer scientist in Auckland, New Zealand, recently posed the following six tasks:

1. Cut a square into three congruent parts.

2. Cut a square into three similar parts, just two of which are congruent.

3. Cut a square into three similar parts, no two congruent.

4. Cut an equilateral triangle into three congruent parts.

5. Cut an equilateral triangle into three similar parts, just two of which are congruent.

6. Cut an equilateral triangle into three similar parts, no two congruent.

The solution to the first task is obvious (see Figure 1). It is surely unique, though I know of no proof. Ian Stewart and A. Womstein have shown that no rectangle can be divided into three congruent polyominoes unless the pieces are rectangles.[1]

Figure 1.

This article first appeared in *Quantum* (May/June 1994).
[1]*Journal of Combinatorial Theory,* Series A, Vol. 61, September 1992, pp. 130–36.

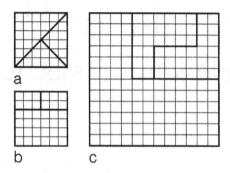

Figure 2.

Figure 2 shows three solutions to the second dissection task.

Task 3 is more difficult. Scherer found the pattern shown in Figure 3. The solution is not unique, because the slanting line can assume an infinity of positions. The one shown may be the one in which line segments have the smallest possible integer lengths.

As mathematician Robert Wainwright of Plainview, New Jersey, has observed, Figure 2b results when the slanting line is orthogonal.

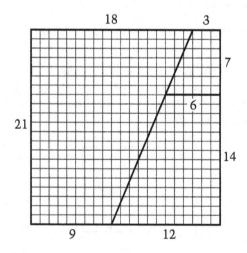

Figure 3.

We turn now to the three equilateral triangle tasks.

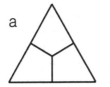 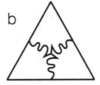

Figure 4.

The fourth task obviously has an infinity of solutions, obtained by rotating the three trisecting lines about the central point (Figure 4). The trisecting lines need not be straight. They can be as wiggly as you like, provided that they are identical and do not intersect (Figure 4b).

Scherer found an elegant solution to the fifth task (Figure 5). It's believed to be unique. Note its similarity to Figure 2c.

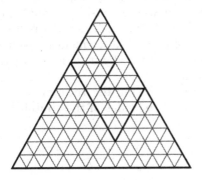

Figure 5.

The sixth task is easily solved (Figure 6). It's probably unique, though no proof is known.

Figure 6.

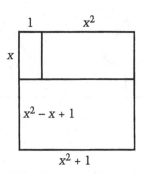

Figure 7.

My only contribution to the six tasks was the rediscovery of a second solution to the third task (Figure 7). I later learned from Scherer that he had found it years earlier. What is the value of x, assuming the smaller side of the smallest rectangle is 1? I thought this would be a simple question to answer. If x isn't rational, surely it's a recognizable irrational, such as 1.732...(the square root of 3), or 1.618...(the golden ratio, often called phi), or some other well-known irrational.

To my amazement, x turned out to be an irrational number I had never encountered before.

The cubic equation relating the ratio of the sides of the smallest rectangle to the ratio of the sides of the similar largest rectangle is

$$\frac{1}{x} = \frac{x^2 - x + 1}{x^2 + 1},$$
$$x^3 - 2x^2 + x - 1 = 0,$$
$$(x^2 - x)(x - 1) = 1.$$

The decimal expansion of x is 1.75487766624669276.... As Wainwright pointed out, the number is closely related to phi, the golden ratio. The reciprocal of phi equals phi minus one. The reciprocal of x equals $(x - 1)^2$. Other equalities are

$$\frac{1}{x^2} = \sqrt{x} - 1; \quad \sqrt{x} = \frac{1}{x - 1}.$$

I propose calling this number "high-phi." Donald Knuth, Stanford University's noted computer scientist, suggested giving it the symbol ⴕ, in which the little circle of phi is raised. He pointed out in a letter how close a modified fraction for high-phi resembles the continued fraction for phi. Phi is the limit of

$$1 + \cfrac{1}{1 + \cfrac{1}{1 + \cfrac{1}{1 + \cfrac{1}{1 + 1_{\cdot_{\cdot_{\cdot}}}}}}}$$

Add square root signs and you get the modified continued fraction for high-phi:

$$1 + \cfrac{1}{\sqrt{1 + \cfrac{1}{\sqrt{1 + \cfrac{1}{\sqrt{1 + \cfrac{1}{\sqrt{1 + 1}_{\cdot_{\cdot_{\cdot}}}}}}}}}}$$

As Knuth wrote, the series converges more rapidly than the series for phi, giving values that are alternately over and under the true value: 1, 2, 1.71, 1.765, 1.753, 1.7554, ... Knuth also called attention to the following equality for high-phi:

$$1 + \frac{1}{\phi - 1} = \phi + \frac{1}{\phi}.$$

Karl Scherer points out that the three rectangles in my figure have areas of x, x^3, and x^4. And if the original square has a side length of 1, the rectangles have areas of $1/x$, $1/x^2$, and $1/x^4$. This shows that $1 = 1/x + 1/x^2 + 1/x^4$, and the ratio of the largest rectangle to the rest of the square is $\sqrt{\phi}$.

Scherer suggests the terms phi-two, phi-three, and so on, for the first terms of the series of solutions for the equation

$$\frac{1}{x} = (x - 1)^n.$$

He conjectures that 1 is the sum of the infinite series of the reciprocals of phi-two, phi-three, phi-four, and so on. In brief,

$$1 = \sum_0^n \phi_n^{-2^n}.$$

Can any reader prove or refute this conjecture? Is it not surprising that such a simple geometrical construction would generate such a curious number? Note that 666, the number of the beast in the Book of Revelations, follows its first six decimal digits.

Postscript

Robert Wainwright found a fourth way to cut a square into three similar parts, just two congruent, as shown in Figure 8. It resembles my third pattern (Figure 2c), but the proportions are different.

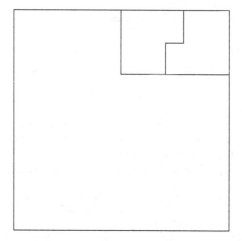

Figure 8.

Rodolfo Marcelo Kurchan, in Buenos Aires, and the Canadian mathematician Andy Liu independently considered the task of cutting a square into four similar parts, meeting the following provisos:

1. All four congruent.

2. Just three congruent.

3. Two congruent, and a different pair also congruent.

4. Two congruent, two not.

5. No two congruent.

They found solutions to all but the fifth task. Does such a pattern exist?

Samuel J. Maltby, in the *Journal of Combinatorial Theory*, Series A (Vol. 66, April 1994, pp. 40–52) proved that if any rectangle (including, of course, the square) is dissected into three congruent parts, the pieces must be rectangles. Still open is whether this is also true of all convex quadrilaterals.

Several readers noticed that my Figure 6 dissection can be easily extended to produce n smaller and smaller similar triangles.

Scherer's square problem was generalized by Byung-Kyu Chun, Andy Liu, and Daniel van Vliet in "Dissecting Squares into Similar Rectangles," in *Crux Mathematicorum* (Vol. 22, 1996, pp. 241–248). They described their generalization as follows:

> We generalize the problem of Scherer and Gardner for the square as follows. Given any integer $m > 1$ and any of its $2^{(m-l)}$ compositions, or ordered partitions, $m = a_1 + a_2 + \ldots + a_n$, dissect a square into m similar pieces so that there are a_1 congruent pieces of the largest size, a_2 congruent pieces of the next largest size, and so on. In the original problem, $m = 3$ and the compositions are: (1) 3; (2a) 1 + 2; (2b) 2 + 1; (3) 1 + 1 + 1.
>
> Our main result is that the dissection problem always admits a solution using rectangular pieces if and only if the composition is not of the form $k + 1$, where k is any positive integer. These solvable cases are covered by two constructions which are only slightly different.

Scherer's conjecture, which I gave at the end of my article, was shot down by Scherer himself.

Ian Stewart's *Scientific American* column for June 1996, and a brief note at the end of his November 1996 column, concerned what he calls the "plastic number." This was the name given to it recently by Richard Padovan, an Italian architect who credited the number's discovery to a French architect in 1924. The number derives from the sequence 1, 1, 1, 2, 2, 3, 4, 5, 7, 9, 12, 16, 21..., in which each number is the sum of the second and third numbers preceding it.

The sequence is geometrically digrammed by the spiraling equilateral triangles shown in Figure 9. Like the familiar spiral of squares that generate the Fibonacci numbers, a logarithmic spiral can be drawn on the pattern. In the case of the whirling squares, the spiral is inscribed. In the case of the whirling triangles, it is circumscribed.

Two adjacent numbers of the Fibonacci sequence have ratios that converge on phi, the golden ratio 1.618 033.... This irrational number is the value of x in the equation $x^2 - x - 1 = 0$. The other sequence is closely related. Ratios of adjacent terms converge on the irrational number 1.324 717 957..., the value of x in the equation $x^3 - x - 1 = 0$.

Midhat J. Gazalé, an Egyptian electrical engineer who retired in 1993 as president of AT&T France, sent me a fascinating unpublished paper on the plastic number and its properties, including the whirling triangles and their circumscribed logarithmic spiral. Gazalé thought the term "plastic number" too ugly (I agree). He rechristened it the "silver number" to honor its close affinity to the golden ratio. I suggest calling the sequence that generates it the silver sequence.

And now for a big surprise. In my Figure 7, the ratio of $x^2 - x + 1$ to x is the silver number! In other words, high phi (1.754...) squared, minus high phi, plus 1, divided by high phi, is the silver number 1.324.... There are even simpler formulas relating the two irrationals:

$$\frac{1}{\varphi - 1} \quad and \quad \varphi^2 - \varphi.$$

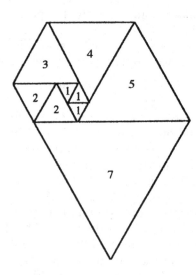

Figure 9.

Chapter 17
Lewis Carroll's Pillow-Problems

Lewis Carroll was the pen name of Charles Dodgson, who taught mathematics at Christ Church, one of the Oxford University colleges in England. He is best known, of course, as the author of two immortal fantasies about Alice and a long nonsense ballad called *The Hunting of the Snark*.

In 1893 Carroll published a little book of seventy-two original mathematical puzzles, many of them not easily solved. The book's title was *Pillow-Problems Thought Out During Sleepless Nights*. For the book's second edition he changed the last two words to "wakeful hours" so readers wouldn't think he suffered from chronic insomnia. A new preface was added to the fourth edition (1895). Carroll intended the book to be part II of what he called *Curiosa Mathematica*. Part I, *A New Theory of Parallels*, was too serious to be called recreational even though it was written with the usual Carrollian humor.

The most interesting puzzles in *Pillow-Problems* concern probability. The first one, problem 5, is simple to state but extremely confusing to analyze correctly:

★ ★ ★

A bag contains one counter, known to be either white or black. A white counter is put in, the bag shaken, and a counter drawn out, which proves to be white. What is now the chance of drawing a white counter?

★ ★ ★

As Carroll writes, one is tempted to answer 1/2. Before the white counter is withdrawn, the bag is assumed to hold with equal probability either one black and one white counter, or two white counters. If the counters in the bag are black and white, a black counter will remain after the white one is taken. If the counters are both white,

This article first appeared in *Quantum* (March/April 1995).

a white counter will remain after a white one is drawn. Because the two states of the bag are equally probable, it seems that after a white counter is taken, the remaining counter will be black or white with equal probability.

Carroll claims correctly that the above argument, though intuitively plausible, is dead wrong. Let A stand for a white counter in the bag at the outset, B for a black counter, and C for the added white counter. After a white counter is taken, there are three, not two, equally possible states:

1. C has been taken, leaving A.

2. A has been taken, leaving C.

3. C has been taken, leaving B.

In the first two cases a white counter remains in the bag. In the third case, the remaining counter is black. The somewhat surprising answer, therefore, is 2/3.

The probability of first drawing a white counter is 3/4, and the probability that the remaining counter is white is also 3/4. Of course, as soon as you see that the counter taken is white, the probabilities alter. If black, the other counter is white with certainty. If white, the other counter is white with a probability of 2/3, and black with a probability of 1/3. All this can be made clear with an inverted tree diagram (see Figure 1).

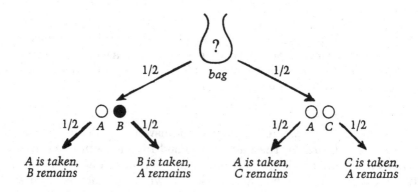

Figure 1.

The fractions represent probabilities. The probability of each of the four outcomes (bottom row) is 1/2 times 1/2, or 1/4. The diagram shows that three times out of four a white counter will be drawn, and three times out of four a white counter remains in the bag. If drawing a black counter is not considered—assume that if this happens the black counter will be replaced and drawing continued until a white counter is taken—the remaining counter is white two times out of three.

The problem is easily modeled with playing cards. Shuffle a deck, spread it face down, and remove a card without looking at its face. Beside it place face down a card you know to be red. Turn your back while a friend mixes the positions of the two cards. Turn around and put a finger on one card. The chance that it's red is 3/4, and the chance the other card is red is also 3/4. Turn over the card you're touching. If it's black, the other card must be red. If it's red, the probability the other card is red goes down to 2/3.

The book's last problem, No. 72, has been the subject of much controversy.

⋆   ⋆   ⋆

> A bag contains 2 counters, as to which nothing is known except that each is either black or white. Ascertain their colours without taking them out of the bag.

⋆   ⋆   ⋆

Here is Carroll's surprising answer:

We know that, if a bag contained 3 counters, 2 being black and one white, the chance of drawing a black one would be 2/3; and that any *other* state of things would *not* give this chance.

Now the chances, that the given bag contains $(\alpha)BB, (\beta)BW, (\gamma)WW$, are respectively 1/4, 1/2, 1/4.

Add a black counter.

Then the chances, that it contains $(\alpha)BBB, (\beta)BWB, (\gamma)WWB$, are, as before, 1/4, 1/2, 1/4.

Hence the chance, of now drawing a black one,

$$= 1/4 \cdot 1 + 1/2 \cdot 2/3 + 1/4 \cdot 1/3 = 2/3.$$

Hence the bag now contains BBW (since any *other* state of things would *not* give this chance).

Hence, before the black counter was added, it contained BW, i.e. one black counter and one white.

The proof is so obviously false that it's hard to comprehend how several top mathematicians could have taken it seriously and cited it as an example of how little Carroll understood probability theory! There is, however, not the slightest doubt that Carroll intended it as a joke. He answered all thirteen of the other probability questions in his book correctly. In the book's introduction he gives the hoax away:

> If any of my readers should feel inclined to reproach me with having worked too uniformly in the region of Commonplace, and with never having ventured to wander out of the beaten tracks, I can proudly point to one Problem in 'Transcendental Probabilities'—a subject in which, I believe, *very* little has yet been done by even the most enterprising of mathematical explorers. To the casual reader it may seem abnormal, and even paradoxical; but I would have such a reader ask himself, candidly, the question "Is not Life itself a Paradox?"

It was characteristic of Carroll that he ended his book with a choice specimen of Carrollian nonsense.

Chapter 18
Lewis Carroll's Word Ladders

Doublet tasks consist of changing one word to another by altering single letters at each step to make a different word. The two words at the beginning and end of such a chain must, of course, be the same length, and they should be related to each other in some obvious way. They must not have identical letters in the same positions. All words in the chain should be common English words, proper names excluded. A "perfect" solution has a number of steps equal to the number of places the given words differ. For example: Cold, Cord, Card, Ward, Warm. If a perfect change is not possible, the best solution is the shortest chain. For playing doublets as a game with two or more players, Lewis Carroll invented a set of scoring rules to determine who wins.

The first mention of the game in Carroll's *Diary* is on March 12, 1878, when Carroll reports teaching "Word Links" (his original name for the game) to guests at a dinner party. He had invented the game, he tells us in a pamphlet, on Christmas Day in 1877 for two little girls who "found nothing to do."

Carroll's hand lettered "Word-Links: A Game for Two Players, or a Round Game" was written in April 1878. Later that year he printed a revised version as a four-page pamphlet. Starting with the March 29, 1879, issue of *Vanity Fair*, Carroll contributed a series of articles on doublets. The first was followed by an article announcing a doublet competition and a third article giving a new method of scoring. In 1879 Macmillan gathered the *Vanity Fair* articles into a 39-page book, with red dot covers, titled *Doublets: A Word Puzzle*. A 1880 second edition was enlarged to 73 pages. *The Lewis Carroll Picture Book* reprints part of this edition. Later that same year Macmillan published a third edition, revised and enlarged to 85 pages.

Carroll took the name doublets from a line of the witches incanta-

---

This article first appeared in *Math Horizons* (November 1994).

tion in Shakespeare's *Macbeth*: "Double, double, toil and trouble"—a line Carroll placed on the title page of his book.

On May 11, 1885, Carroll mentions in his *Diary* that he has extended his list of seven-letter word pairs that can be linked together to more than 500.

Doublets became a parlor craze in London, and has been a much practiced form of word play ever since. They have been called by other names, such as "word ladders," and (in Vladimir Nabokov's novel *Pale Fire*) "word golf." Enormous energies have been expended on finding shortest ladders for a given pair of words. Computer software containing all English words is now obtainable, and programs have been written for finding minimum chains in just a few seconds. The task is equivalent to finding the shortest routes connecting two points on a graph.

Donald Knuth, Stanford University's noted computer scientist, has constructed a graph on which 5,757 of the most common five-letter English words (proper nouns excluded) are represented by points, each joined by a line to every word to which it can be changed by altering just one letter. The graph has 14,135 lines. Once in a computer's memory, programs can be written that will determine in a split second the shortest word ladder joining any two words on the graph. Knuth found three-letter words too simple, and six letter words less interesting because not too many can be connected.

Most pairs of five-letter words on Knuth's list can be joined by ladders. Some—Knuth calls them "aloof" words because one of them is the word aloof—have no neighbors. The graph has 671 aloof words, such as earth, ocean, below, sugar, laugh, first, third, ninth. Two words, bares and cores, are connected to 25 other words; none to a higher number. There are 103 word pairs with no neighbors except each other, such as odium-opium, and monad-gonad. Knuth's 1992 Christmas card featured the smallest ladder (eleven steps) that changes sword to peace by using only words found in the Bible's Revised Standard Version.

I have taken the above information from the eight pages devoted to doublets in the first chapter of Knuth's book *The Stanford GraphBase* (Addison-Wesley, 1993). Knuth will cover the topic more fully in his forthcoming three-volume work on combinatorics in his classic "Art of Computer Programming" series. For hints on how to solve doublets without a computer see his article "Lewis Carroll's Word, Ward, Ware, Dare, Dame, Game," in *Games* magazine (July–August, 1978).

It has been pointed out that doublets resemble the way in which evolution creates a new species by making small random changes in the "genes" that are intervals along the helical DNA molecule. Carroll himself, although a skeptic of Darwin's theory, evolved Man from Ape in six steps:

APE ARE ERE ERR EAR MAR MAN

When I gave this solution in a *Scientific American* column on Mathematical Games (the column is reprinted as Chapter 4 in my *New Mathematical Diversions*), two readers produced a shorter solution:

APE APT OPT OAT MAT MAN

In a letter of March 12, 1892, (See Morton Cohen's *The Letters of Lewis Carroll*, Volume 2, page 896), Carroll added a rule that allows one to rearrange the letters of any word, counting this as a step. With such increased freedom, he pointed out, many impossible doublets, such as changing Iron to Lead, can be achieved:

IRON ICON COIN CORN CORD LORD LOAD LEAD

It is difficult but not impossible for a word chain to form a sentence. In *Vanity Fair* (July 26, 1879), one of Carroll's doublets asked "WHY is it better NOT to marry?" to change WHY to NOT he added this proviso: "the chain made [WHY to NOT]...should embody the following observation: that lovers, during the temporary insanity of courtship, too often fail to recognize the grave prudential reasons which should deter them from taking this fatal step." Here is Carroll's clever solution:

WHY WHO WOO WOT NOT

The mathematician and writer of science fiction, Rudy Rucker, has likened doublets to a formal system. The first word is the given "axiom." The steps obey "transformation rules" and the final word is the "theorem." One seeks to "prove" the theorem by the shortest set of transformations.

Many papers on doublets have appeared in the journal *Word Ways*, a quarterly devoted to linguistic amusements. An article in the February 1979 issue explored chains that reverse a word, such as TRAM to MART, FLOG to GOLF, LOOPS to SPOOL, and so on. The author asks if an example can be found using a six-letter word.

Is there a closed chain, I wonder, that changes SPRING to SUMMER to AUTUMN to WINTER, then back to SPRING? If so, what is the shortest solution?

A.K. Dewdeny, in a Computer Recreations column in *Scientific American* (August 1987), calls a graph connecting all words of n letters, a "word web." He shows that all 2-letter words are easily joined by such a web, and asks if anyone can construct a complete word web for 3-letter words.

ROGUE	QUELL	KETTLE	COSTS	SHOES
vogue	quill	settle	posts	sloes
vague	quilt	settee	pests	floes
value	guilt	setter	tests	floss
valve	guile	better	tents	gloss
halve	guide	betted	tenth	glass
helve	glide	belted	tench	class
heave	glade	bolted	teach	crass
leave	grade	bolter	peach	cress
lease	grave	bolder	peace	crest
least	brave	bolder	peace	CRUST
BEAST	BRAVO	HOLDER	PENCE	

BLACK	BEANS	GRASS	STEAL	WHEAT
clack	beams	crass	steel	cheat
crack	seams	cress	steer	cheap
track	shams	tress	sheer	cheep
trick	shame	trees	shier	creep
trice	shale	frees	shies	creed
trite	shall	freed	shins	breed
write	shell	greed	chins	BREAD
WHITE	SHELF	GREEN	COINS	

FURIES	TEARS	PITCH	FLOUR	RAVEN
buries	sears	pinch	floor	riven
buried	stars	winch	flood	river
burked	stare	wench	blood	riser
barked	stale	tench	brood	MISER
barred	stile	tenth	broad	
BARREL	SMILE	TENTS	BREAD	

Table 1. These are Carrell's best solution to fifteen doublets. Can any be improved?

Postscript

My asking for a closed chain of the seasons surely has no solution. Donald Knuth pointed out that AUTUMN has no neighbor. He suggested adopting Carroll's broader rules that allow anagramming a word after a letter chain. MUTUAL, for instance, then becomes a neighbor of AUTUMN. Even with these new rules I still don't know if a closed chain exists.

Here is how Knuth improved some of Carroll's solutions for the doublets that I gave:

SHOES	BLACK	COSTS	COSTS	GRASS	GRASS
shops	brack	coots	coats	gross	crass
chops	brace	clots	boats	grows	cress
crops	trace	plots	blats	grown	tress
cross	trice	plats	plats	groat	trees
cress	trite	plate	plate	great	treed
crest	write	place	place	greet	greed
CRUST	WHITE	peace	peace	GREEN	GREEN
		PENCE	PENCE		

Chapter 19
The Ant on 1 x 1 x 2

An ant is at corner A of a 1 × 1 × 2 box. It crawls along the surface along a geodesic, the shortest possible path, to a point B. Where is B located to make the path as long as possible?

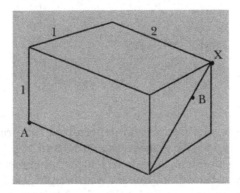

Figure 1. A 1 × 1 × 2 bicube, a solid formed by joining two cubes.

Intuitively one would guess B to be at the corner marked X because this is the point the farthest from corner A. Yoshiyuki Kotani, a professor of mathematics in Saitama, Japan, recently made a surprising discovery. Point B is not at X, but one-fourth of the way down the diagonal of the square face as shown!

The Geodesic from A to X is easily traced by unfolding the solid along a hinged edge as shown in Figure 2. The Pythagorean theorem gives the path as the square root of 8, or 2.828.... If you trace the geodesic from A to B, the ant can take either of the two routes shown in Figures 3 and 4. By symmetry there are two similar routes along the hidden sides of the solid. Two of the four paths go over two sides, and

---

This article first appeared in *Math Horizons* (February 1996).

the other two go over three sides. Applying the Pythagorean theorem
to these four paths, they all have the same length of 2.850..., about
0.022 longer that the path from A to X!

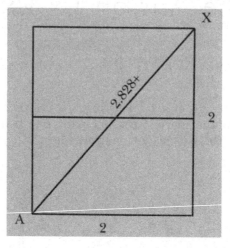

Figure 2.

I do not know whether Kotani generalized the problem to $1 \times 1 \times n$ solids. In any case, physicist Richard Hess, who first called my
attention to the problem, and four mathematicians to whom I sent
the problem (Ken Knowlton, Robert Wainwright, Dana Richards, and
Brian Kennedy) each independently solved this more general case. I
expected that calculus would be required, but it turns out that by
unfolding the solid along hinged edges, and applying basic algebra,
the formula for the location of B is not too difficult to find.

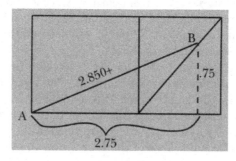

Figure 3.

Label the diagonal of the square face with 0 at X and 1 at the diagonal's bottom corner. The distance of B from X, along this diagonal, as a function of n, is $(n-1)/2n$. If $n = 2$, the formula puts B a quarter of the way down the diagonal. If $n = 3$, B is 1/3 of the way down. As n approaches infinity, B approaches 1/2 at the limit, placing it at the center of the diagonal. Of course n can take any real value greater than 1 and not necessarily an integer.

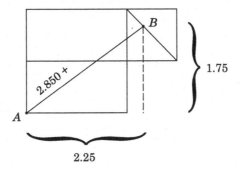

Figure 4.

Problems about a spider at spot A on the wall of a room that crawls along a geodesic to catch a fly at spot B are in many classic puzzle books. Henry Dudeney, the British puzzle maker, gives such a problem in *The Canterbury Puzzles*, and his American counterpart Sam Loyd has the same problem in his *Cyclopedia of Puzzles* (page 219). The French mathematician Maurice Kraitchik poses a similar puzzle in *Mathematical Recreations*, with an illustration showing all ways of unfolding the room.

Such problems can be further generalized to solids (or rooms) of dimensions $a \times b \times c$. A more difficult question, suggested by computer scientist Donald Knuth, is to find maximum-length geodesics on such solids. For example, the maximum geodesic on the $1 \times 1 \times 2$ solid is *not* from the center of one square face to the center of the other, a distance of 3. Hess has made the surprising discovery that the maximum path has a length of slightly more than 3.01. But Hess's results on maximum-length geodesics, as yet unpublished, are a long story.

Chapter 20
Three-Point Tiling

Some tiling problems are infuriating because they look as if they have easy solutions but then turn out to be enormously difficult. Two good examples are two closely related problems involving the tiling of points arranged in the form of equilateral triangles with n points on the side.

The Greeks used such lattices to diagram what they called triangular numbers—numbers that are partial sums of 1, 2, 3, ... If the number of points is even, it is easy to "tile" any triangle, without overlap, by line segments that join two adjacent dots.

Two hard questions now arise.

If the number of points is a multiple of 3, then
a) can the triangle be tiled with 3-point straight lines?
b) can it be tiled with little triangles consisting of three mutual adjacent points?

To put the same two questions in a different form, can equilateral triangles formed by adjacent hexagons—when the number of hexagons is a multiple of 3—be tiled with either of the two "trihexes" (as I called them in one of my *Scientific American* columns), the straight trihex and the triangular trihex? We will consider both tasks in the point form rather than the polyhex form.

Any triangle, whose points are a multiple of 3, can be tiled with a mixture of straight lines and triangles by adopting the "canonical" pattern shown in Figure 1. But can such triangles be tiled with either tile alone?

I do not know the origin of either of these two seemingly easy problems, but as far as I am aware, they were first answered in a pioneering paper by J.H. Conway and J.C. Lagarias, "Tiling with Polyominoes and Combinatorial Group Theory," in the *Journal of Combinatorial Theory*; Series A, Vol. 53 (1990) pp. 183–208. The authors

This article first appeared in *Cubism for Fun* (No. 28, April 1992).

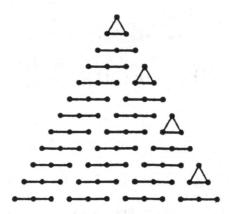

Figure 1. Canonical tiling for $n = 12$.

were able to prove that *no* triangle can be tiled with 3-point lines, and that they can be tiled with 3-point triangles if and only if $n = 0, 2, 9$, or 11 (modulo 12). Both proofs are long and difficult, making use of what the authors call "boundary invariants". These are combinatorial group invariants associated with the boundaries of the tiles and the boundaries of the regions to be tiled. The authors admit that their proofs are "somewhat complicated, and it is reasonable to ask if simpler solutions exist".

Neither problem, they also maintain, can be solved by a coloring procedure. This may be true, but Solomon Golomb, at the University of Southern California, found a clever coloring proof for the linear tiles that proves impossibility for all values of n, except for $n = 0$ or -1 (modulo 9). His proof first appeared in his column "Golomb's Gambits", in the *John Hopkins Magazine*, October 1990. It goes as follows: Label the rows of the triangle with three colors, a, b, c, that repeat cyclically as shown in Figure 2 on the $n = 11$ triangle.

Each vertical tile (slanting either way) covers one point of each color. Each horizontal line covers three points of the same color. Clearly, if we add the number of covered spots of each color, any pair of sums must differ by a multiple of 3. However, when we count the number of spots of each color they do not differ from one another by multiples of 3. This is a clear contradiction that proves impossibility. Unfortunately, the proof fails when $n = 0$ or -1 (modulo 9).

A similar proof of my own is based on labeling the rows 1, 2, 3. Because each vertical tile has a sum of 6, and each horizontal tile has

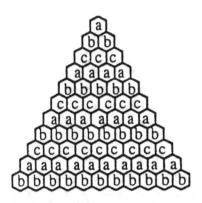

Figure 2. Coloring pattern for Golomb's proof.

a sum of 3, 6, or 9, the sum of all the tiles must be a multiple of 3, except when $n = 0$ or -1 (modulo 9). For a while I thought I had a simple impossibility proof for all linear tiles, but it was quickly and independently shot down by Fan Chung and Herbert Taylor.

A quite different proof that applies to both tiling tasks has been discovered by Donald C. West, at SUNY, Plattsburgh, NY, and may appear in a forthcoming paper. It, too, is neither short nor simple, resting on a way of transforming patterns to the canonical form.

Figure 3 shows how an n-triangle can be tiled with the little triangles if $n = 9$ or 11. Both patterns have a pleasant three-fold rotational symmetry. The pattern for $n = 9$ is unique. The $n = 11$ triangle also has an unsymmetric tiling, however, which is obtained by a 180-

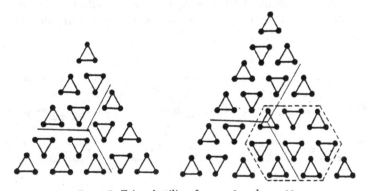

Figure 3. Triangle tiling for $n = 9$ and $n = 11$.

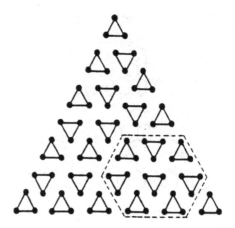

Figure 4. Triangle tiling for $n = 12$.

degree rotation of a hexagonal region of 8 small triangles (indicated by dotted lines). This tiling can also be obtained by adding a bottom row of triangles (alternating in orientation) to the $n = 9$ tiling.

In Figure 4 we show a tiling for the $n = 12$ triangle, obtained by adding two rows of triangles to the $n = 9$ tiling. Another tiling may be obtained from this one by a 180-degree rotation of a hexagonal region equal to the one in the $n = 11$ triangle. The number of patterns for the $n = 12$ triangle, and any larger triangle, is unknown.

Michael Beeler, from Newton, MA, found a general way to tile any large N-triangle, $N = 0, 2, 9, 11$ (modulo 12) with small triangular tiles. For both problems, the crucial lemma proves the existence of tileable strips of width 6 and of arbitrary length, at least 2. For the present triangle problem these strips are constructed as follows: First two types a and b of building blocks, each consisting of two small triangles, are defined (see Figure 5).

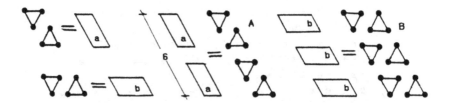

Figure 5. Beeler's method of tiling strips of width 6.

Two blocks a form a strip A of width 6 and length 2. Three blocks b form a strip B of width 6 and length 3. Any strip of width 6, and of even length, can be filled by strips A. If a 6-wide strip has odd length, a strip B is taken off and a strip of even length remains, which can be filled by strips A. Now, all ingredients needed to tile a large N-triangle are collected. Figure 6 shows the tiling plan, which makes use of a combination of 6-wide strips (according to Figure 5), side-12 triangles (according to Figure 4) and one triangle of side 2, 9, or 11 (according to Figure 3).

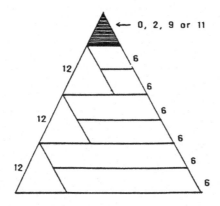

Figure 6. Tiling plan for large $N = 0, 2, 9, 11$ [mod 12] triangles.

Questions left over:

- Is there a simpler way of proving the Conway/Lagarias results?

- What is the number of different triangle tilings for large N?

- What is the number of rotational-symmetric triangle tilings?

Chapter 21
Lucky Numbers and 2187

The house where I grew up as a child in Tulsa, Oklahoma, has an address of 2187 S. Owasso. Of course I never forgot this number. Many years ago, when I was visiting my imaginary friend Dr. Irving Joshua Matrix, the world's most famous numerologist, I asked him if there was anything remarkable about 2187.

He immediately replied: "It is 3 raised to the power of 7. If you write it in base 3 notation it is 10,000,000."

"I'm amazed you would know that!" I exclaimed. "Anything else unusual about 2187?"

"My dear chap," Dr. Matrix responded with a heavy sigh, "*every* number has endless unusual properties. Exchange the last two digits of 2187 to make 2178, multiply by 4, and you get 8712, the second number backward. Take 2187 from 9999 and the result is 7812, its reversal. Multiplying 21 by 87 produces 1827, the same digits in a different order. They are in correct order in the product of 27 and 81. And have you noticed that the first four digits of the constant e, 2, 7, 1, 8, and the number of cubic inches in a cubic foot, $12^3 = 1728$, are each permutations of 2187? You might ask your readers how quickly they can insert plus or minus signs inside 2187 to make the expression add to zero."

I was struggling to jot all this down on my notepad when Dr. Matrix added: "And 2187 is, of course, one of the lucky numbers."

I had never heard of lucky numbers. What follows is a summary of what I learned about them from Dr. Matrix, and from the references listed at the end of this article.

The notion of lucky numbers originated about 1955 with Stanislaw Ulam, the great Polish mathematician who co-invented the H-bomb and was the father of cellular automata theory. It is one of the most studied types of what are called "sieve numbers." The oldest, most

This article first appeared in *The Mathematical Intelligencer* (Vol. 19, No. 2, 1997).

important sieve numbers are the primes. They are called sieve numbers because they can be generated by what is known as the Sieve of Eratosthenes.

★   ★   ★

Imagine all the positive integers written in counting order. Cross out all multiples of 2, except 2. The next uncrossed-out number is 3. Cross out all multiples of 3, except 3. Continue in this way, sieving out multiples of 5, 7, 11, and so on. The numbers that remain (except for the special case of 1) are the primes.

★   ★   ★

The sieving process is slow and tedious, but if continued to infinity it will identify every prime.

Using a sieve for generating lucky numbers is similar. Curiously, it produces numbers closely related to primes even though they are mixtures of primes and composites (non-primes). Here is how the procedure works.

Step 1: Cross out every second number: 2, 4, 6, 8, . . . , leaving only the odd integers.

Step 2: Note that the second uncrossed-out integer is 3. Cross out every third number not yet eliminated: 5, 11, 17, 23, ...

Step 3: The third surviving number from the left is 7. Cross out every seventh integer not yet crossed out: 19, 39, ...

Step 4: The fourth number from the left is 9. Cross out every ninth number not yet eliminated, starting with 27.

As you continue in this fashion you will see that certain integers permanently escape getting killed. Ulam called them "lucky numbers." Figure 1 lists all the luckies less than 1,000.

Eratosthenes's sieve abolished all numbers except the primes. The procedure is based on division. Ulam's sieve, on the contrary, is based entirely on a number's *position* in the counting series. Using Eratosthenes's sieve you have to count every integer as you go along. Using Ulam's sieve you count only the integers not previously eliminated.

Although the luckies are identified by a sieving process completely different from Eratosthenes's sieve, the amazing thing is that luckies share many properties with primes. The density of luckies in a given interval among the counting numbers is extremely close to the density of primes in the same interval. For example, there are 25 primes less

```
 1   3   7   9  13  15  21  25  31  33  37  43  49  51  63
67  69  73  75  79  87  93  99 105 111 115 127 129 133 135
141 151 159 163 169 171 189 193 195 201 205 211 219 223 231
235 237 241 259 261 267 273 283 285 289 297 303 307 319 321
327 331 339 349 357 361 367 385 391 393 399 409 415 421 427
429 433 451 463 475 477 483 487 489 495 511 517 519 529 535
537 541 553 559 577 579 583 591 601 613 615 619 621 631 639
643 645 651 655 673 679 685 693 699 717 723 727 729 735 739
741 745 769 777 781 787 801 805 819 823 831 841 855 867 873
883 885 895 897 903 925 927 931 933 937 957 961 975 979 981
991 993 997
```

Figure 1. A computer printout of lucky numbers less than 1,000, supplied by Charles Ashbacher. Note that '99 will be a lucky year.

than 100, and 23 luckies less than 100. The overall asymptotic density for each type of number is the same!

The distances between successive primes and the distances between successive luckies keep growing longer as the numbers grow in size. These distances also are almost the same for both number types. The number of twin primes—primes that differ by 2—is close to the number of twin luckies. There are eight twin primes less than 100, and seven twin luckies in the same interval. Although primes play a much more significant role in number theory than luckies, the similarities suggest that many of the properties of primes are less unique than previously assumed. Their properties may be more a product of sieving than anything else!

The most notorious unsolved problem involving primes, now that Fermat's Last Theorem has been proved, is the Goldbach conjecture. It states that every even number greater than 2 is the sum of two primes. There is a similar unsolved conjecture about luckies: that every even number is the sum of two luckies. This has been computer tested for integers up to 100,000, and perhaps further than that, in recent years, without finding an exception.

In a 1996 booklet about number problems, Charles Ashbacher, of Cedar Rapids, Iowa, conjectures that every lucky number appears at the tail of a larger lucky. For example, 7 is at the end of 37; 9 at the end of 49; 15 at the end of 615; and so so on. Lucky 87 is at the end of my old house number 2187. Lucky 579 is at the end of lucky 96579. Ashbacher wrote a computer program that verified his conjecture for 22 of the first 100 luckies. This suggests, he writes, that his conjecture is a good bet.

It is easy to determine if certain large numbers are not lucky. Consider 98765. We can quickly tell it is not lucky because it has a digital root of 8. The digital root of a number is its equivalence modulo 9—that is, the remainder when divided by 9. If there is no remainder, the digital root is 9. A digital root is quickly obtained by adding the digits of a number, then adding again if the sum has more than one digit, and continuing this way until just one digit remains. Lucky numbers display all digital roots except 2, 5, and 8. Why? Because all numbers with those three digital roots have the form $3k + 2$ and the first two sieving steps eliminate them all.

Dr. Matrix called my attention to the curious fact that 13, considered the most unlucky of all numbers, is the fifth lucky, the sixth prime, and the seventh Fibonacci number.

A few weeks after my meeting with Dr. Matrix I received from him a fax message listing the following identities:

2187 + 1234 = 3421
2187 + 12345 = 14532
2187 + 123456 = 125643
2187 + 1234567 = 1236754
2187 + 12345678 = 12347865
2187 + 123456789 = 123458976

Note how the sums on the right are permutations of the numbers added to 2187.

It has been proved that no polynomial formula will generate only primes, and I would guess that the same is true for the luckies. However, simple quadratic formulas will generate sequences of primes and luckies. One way to search for such formulas was invented by Ulam. On a square grid write the integers in a spiral fashion as shown in Figure 2, and indicate the luckies by color. Note that nine luckies clump along a diagonal. Applying the calculus of finite differences to these luckies we discover that they are generated by $4x^2 + 2x + 1$, as x takes the values $-3, -2, -1, 0, 1, 2, 3, 4, 5$.

The spiral can start with any higher number to reveal clumps along different diagonals. Leonhard Euler found that $x^2 + x + 41$ generates forty primes by letting x take values 0 through 39. If you write the integers in a spiral starting with 41, these primes will fill the entire diagonal of a 40×40 grid! Is there a quadratic formula equally rich, or perhaps even richer, in finding a clump of luckies? I will be interested in hearing from any reader who finds such a formula.

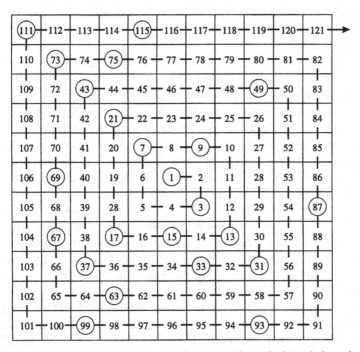

Figure 2. Ulam's spiral technique for finding quadratic lucky-rich formulas.

There is a classic proof by Euclid that there is an infinity of primes. Although it is easy to show there is also an infinity of lucky numbers, the question of whether an infinite number of luckies are primes remains, as far as I know, unproved. Also unsolved is whether there is an infinity of twin luckies.

Dr. Matrix enjoys practical jokes. When we talked about 2187 he pointed out that if this number is divided by 9999 the quotient is 0.218721872187.... I was momentarily surprised until I realized that *any* integer of n digits, not made entirely of nines, when divided by a number consisting of n nines, produces a decimal fraction in which the original number is repeated endlessly as the quotient's period.

"Ulam discovered lucky numbers with his lucky imagination," Dr. Matrix added. "Note the letters at positions 2, 1, 8, and 7 in LUCKY IMAGINATION. What do they spell?"

The first three lucky numbers are 1, 3, and 7. Now 137 not only is a prime but it is one of the most interesting of all three-digit numbers. It is, of course, the notorious fine-structure constant, the most mysterious of all constants in physics. I mentioned this to Dr. Matrix. This prompted him to talk for twenty minutes about 137. Here are some highlights of what he said:

Check the King James Bible's first chapter, third verse, and seventh word. The word is "light." Dr. Matrix reminded me that the fine-structure constant is intimately connected with light.

The reciprocal of 137, or 1 divided by 137, produces the decimal fraction 0072992700072992700.... The period is a palindrome!

Partition 137 into 13 and 7. The thirteenth letter of the alphabet is M and the seventh is G—my two initials!

Chlorophyll, which takes light from the sun to give energy to plants, is made of exactly 137 atoms.

Dr. Matrix asked me to write my old house number twice, 21872187, and put the number into my hand calculator. This number, he informed me, is exactly divisible by 137. I performed the division and sure enough, the read-out displayed the integer 159651. I got an even greater surprise when Dr. Matrix asked me to turn the calculator up-side down. The number was the same inverted!

Dr. Matrix next asked me to divide 159651 by 73. The result was 2187! I later discovered that this was another of Dr. Matrix's hoaxes. Any number of the form of ABCDABCD is evenly divisible by 137 and 73. The reason? ABCDABCD is the product of ABCD and 10001. The two prime factors of 10001 are 137 and 73, so dividing ABCABCD by those two numbers will naturally restore ABCD. Of course the quotient after the first division is not likely to be invertible.

"Is there any connection," I asked, "between the lucky numbers and 666, the famous number of the Beast in New Testament prophecy?"

Dr. Matrix put his fingertips together and closed his emerald eyes for a full minute before he spoke.

"Consider your old house number 2187, and the first four luckies 1, 3, 7, 9. Omit the 1 in each number to leave 287 and 379. Add the two numbers and you get 666. By the way, I forgot to mention earlier that if you divide 18, the middle digits of 2187, by 27, the first and last digits, the quotient is 0.66666666...."

I conclude with a mind-reading trick of my own that involves 2187. Ask someone to put this number into a calculator's display. With your back turned, tell him to multiply it by any number he likes without revealing this number to you. He next calls out, in any order, each digit in the product except one nonzero digit. You at once name the missing digit.

How do you do it? As he calls out digits, keep adding them in your head until you know the digital root of their sum. This is easily done by casting out nines as explained earlier. If the digital root is 9, he omitted 9. If less than 9, he left out a digit equal to the difference between 9 and the digital root. For example, if the digital root is 2, he omitted 7.

I leave it to you to figure out why this always works with 2187. Hint: 2187 has a digital root of 9.

References

[1] "On Certain Sequences of Integers Defined by Sieves." Verna Gardiner, R. Lazarus, Nicholas Metropolis, and Stanislaw Ulam. *Mathematics Magazine*, Vol. 31, 1956, pp. 117–122.

[2] "The Lucky Number Theorem." David Hawkins and W. W. Briggs. *Mathematics Magazine*, Vol. 31, 1957–58, pp. 81–84, 277–280.

[3] *A Collection of Mathematical Problems*. Stanislaw Ulam. (Interscience, 1960), p. 120.

[4] "Prime-like Sequences Generated by a Sieve Process." W. E. Briggs. *Duke Mathematical Journal*, Vol. 30, 1963, pp. 297–312.

[5] "Sieve-generated Sequences." Marvin Wunderlich. *The Canadian Journal of Mathematics*, Vol. 18. 1966, pp. 291–299.

[6] "A General Class of Sieve-generated Sequences." Marvin Wunderlich, *Acta Arithmetica*, Vol. 16, 1969, pp. 41–56.

[7] *Excursions in Number Theory*. C. Stanley Ogilvy and John T. Anderson. (Dover, 1988), pp.100–102.

[8] *The Penguin Dictionary of Curious and Interesting Numbers*. David Wells. (Penguin, 1986), pp. 13, 32–33.

[9] *Unsolved Problems in Number Theory*. Richard K. Guy. (Second edition, Springer-Verlag, 1994), pp. 108–109.

[10] *Zero to Lazy Eight*. Alexander Humez, et al. (Simon and Schuster, 1993), pp. 198–200.

[11] *Collection of Problems on Smarandache Notions*. Charles Ashbacher. (Erhus University Press, 1996), pp. 51–52.

Postscript

My numerology involving 2187 inspired several readers to find other curiosities involving the number.

Monte Zerger discovered that $1^2 + 2^2 + \ldots 21872^2 = 3481976250$, a number containing each of the ten digits! He also pointed out that in *Which Way Did the Bicycle Go?* (1996), the authors Joseph Konhauser, Dan Velleman, and Stan Wagon prove on page 134 (Problem 82) that 2178 is the only four-digit number that reverses when multiplied by 4.

Charles Ashbacher provided a 3×3 magic square, made entirely with lucky numbers; that has the lowest mathematical sum.

195	9	111
21	105	189
99	201	15

The square's magic constant is 315.

Owen O'Shea, in Ireland, noticed that $87^2 - 21^2 = 7128$. He also found the following result: $21 + 87 = 108$, and $87 - 21 = 66$. Multiply 108 by 66 and you get 7128.

These from Dr. Matrix:

Write down all two-digit numbers that can be made with 2, 1, 8, and 7, without duplicating any digit. They add to 594. Multiply by 3 and you get 1782.

Add the fourth powers of 2, 1, 8, and 7. The sum is 6514. Now add the fourth powers of 6, 5, 1, and 4. The result is 2178.

Chapter 22
3 x 3 Magic Squares

Perhaps the oldest of all combinatorial problems about numbers is the task of placing the first nine counting numbers in a 3 × 3 matrix so that each row, column, and main diagonal has the same sum. It turns out that, not counting rotations and reflections as different, there is only one solution:

2	9	4
7	5	3
6	1	8

Legend has it that in the 23rd century B.C. the mythical King Yu saw the pattern as spots on the back of a sacred turtle in the River Lo. Many modern scholars doubt that *lo shu*, the most common name in China for the magic square, is that ancient. They believe that the pattern is not older than the tenth century A.D. At any rate, the name means Lo River writing. The Chinese identify it with their familiar yin-yang circle of light and dark regions. The four even digits (shown shaded above) are identified with the dark yin. The Greek cross made of the five odd digits is identified with the light yang. For centuries the *lo shu* has been used as a charm on jewelry and other objects. Today large passenger ships frequently arrange their shuffleboard numbers in the *lo shu* pattern.

How can we prove that the pattern is unique? The simplest way known to me is first to note that the nine digits add to 45. If each of the three rows (or columns) has the same sum, then the sum must be one third of 45, or 15. We next list the eight possible triads of digits that add to 15:

This article first appeared in *Quantum* (January/February 1996).

$$9 + 5 + 1$$
$$9 + 4 + 2$$
$$8 + 6 + 1$$
$$8 + 5 + 2$$
$$8 + 4 + 3$$
$$7 + 6 + 2$$
$$7 + 5 + 3$$
$$6 + 5 + 4$$

The square's center digit belongs to four lines of three. Five is the only digit to appear in four of the triads; therefore, 5 must go in the center. Digit 9 is in only two triads, so it must go in a side cell, and the digit at the other end of the same line must of course be 1. Digits 3 and 7 are also in just two triads. For symmetry reasons it does not matter how they are placed in the other two side cells. This gives us

	9	
7	5	3
	1	

The remaining empty cells can now be filled with even digits in only one way to complete the magic.

Various mnemonic methods allow one to form the *lo shu* without having to memorize the entire pattern. For me the easiest is to place the even digits in the corners, in sequence from left to right, top to bottom. The odd digits, also in sequence starting with 9 and going backward, form the zigzag pattern show below:

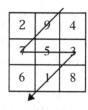

Note that 1, 2, 3 and 7, 8, 9 mark the corners of two isosceles triangles, with 4, 5, 6 along a main diagonal.

In *The Mathematical Gazette* (December 1970, p. 376) R. Holmes pointed out a surprising property of the *lo shu*. Take each row, column, and diagonal (including the four "broken" diagonals) as a three-digit number to be read both forward and backward. The following identities hold:

orthogonals
$$(294 + 753 + 618)^2 = (492 + 357 + 816)^2$$
$$(276 + 951 + 438)^2 = (672 + 159 + 834)^2$$

diagonals
$$(654 + 132 + 879)^2 = (456 + 231 + 978)^2$$
$$(852 + 174 + 639)^2 = (258 + 471 + 936)^2$$

The identities are unaffected if the middle digits of the numbers are deleted, or any two corresponding two digits.

Imagine the *lo shu* to be toroidally connected—that is, imagine the pattern wrapped around both vertically and diagonally, as it would be if drawn on a torus divided into nine cells. I was amazed to discover years ago that if you add the four digits in each 2×2 square, the sums are the nine consecutive numbers from 16 through 24.

Of course, an infinity of 3×3 magic squares can be constructed with other numbers, not necessarily in counting sequence, and including all real numbers. To exclude trivial examples, we assume that no two numbers are alike. For the matrix to be magic, however, the nine numbers must fall into three triads, in each of which the numbers are in the same arithmetic progression (that is, with the same differences). Moreover, the smallest numbers of the three triads must also be in arithmetic progression, though not necessarily with the same differences as the triads. We can express these rules algebraically as follows:

$a + x$	$a + 2y + 2x$	$a + y$
$a + 2y$	$a + y + x$	$a + 2x$
$a + y + 2x$	a	$a + 2y + x$

Note that each line of three cells has a sum of $3a+3y+3x$, or $3(a+y+x)$, proving that the magic constant must be a multiple of 3, and that the center number is one third of the constant.

With the square's algebraic structure in mind, it is possible to construct a fascinating variety of magic squares based on given restraints.

The box below shows a sampling of such squares. Each is the simplest square meeting the restraints, defining "simplest" as the square with the lowest magic constant. Do you see why the number 2 cannot be used in a magic square of primes?

c=27

15	1	11
5	9	13
7	17	3

odd integers

c=30

16	2	12
6	10	13
8	18	4

*even integers
without zero*

c=24

14	0	10
4	8	12
6	16	2

*even integers
with zero*

c=111

67	1	43
13	37	61
31	73	7

*primes
including 1*

c=177

71	89	17
5	59	113
101	29	47

*primes
excluding 1*

c=3177

1669	199	1249
619	1039	1459
829	1879	409

*primes in
arithmetic progression*

c=354

121	114	119
116	118	120
117	122	115

*consecutive
composites*

c=54

27	6	21
12	18	24
15	30	9

*composites in
arithmetic progression*

c=636

222	101	313
303	212	121
111	323	202

*all numbers and c
are palindromes*

Three-by-three magic squares of special types with lowest constant (c). The palindrome square was constructed by Rudolph Ondrejka of Linwood, New Jersey.

In 1987 I offered $100 to anyone who found a 3 × 3 magic square made with consecutive primes. The prize was won by Harry Nelson of Lawrence Livermore Laboratories. He used a Cray computer to produce the following simplest such square:

1,480,028,201	1,480,028,129	1,480,028,183
1,480,028,153	1,480,028,171	1,480,028,189
1,480,028,159	1,480,028,213	1,480,028,141

Martin LaBar, in *The College Mathematics Journal* (January 1984, p. 69), asked if a 3 × 3 magic square exists with nine distinct square numbers. (Such squares exist with eight distinct squares plus a zero.) Neither such a square nor a proof of impossibility has been found. Nelson believes it exists, but beyond the reach of any of today's supercomputers running a reasonable amount of time.

I here offer $100 to the first person to construct such a square. If it exists, its numbers are sure to be monstrously large. John Robertson has shown that the task is equivalent to finding an elliptic curve of the form $y^2 = x^3 - n^2 x$ with three rational points, each the double of another rational point on the curve, having x-coordinates in arithmetic progression.

Henry Ernest Dudeney, in *Amusements in Mathematics* (pp. 124–25) and in his article on magic squares in the fourteenth edition of the *Encyclopaedia Britannica*, defines magic squares based on subtraction, multiplication, and division, and gives 3 × 3 examples of each kind. There are also 3 × 3 *antimagic* squares with the property that no two sums of three lines are alike. If certain provisos are met, a variety of interesting antimagic combinatorial problems result. These variants, however, will have to be the topic of another article.

Surely the most fantastic 3 × 3 magic square ever discovered is one constructed by Lee Sallows, a British electronics engineer who works for the University of Nijmegen in Holland:

5	22	18
28	15	2
12	8	25

It would be hard to guess its amazing property. For each cell, count the number of letters in the English word for its number, then place these counting numbers in the corresponding cell of another 3×3 matrix. For example, "five" has four letters, so 4 goes into the top left corner of the new matrix. Here is the result:

4	9	8
11	7	3
6	5	10

Not only is it another magic square, but its integers are in consecutive order! Sallows calls the first square the *li shu* (*li* for his first name Lee), and the second square its alphamagic partner. His computer investigations of alphamagic squares in more than twenty languages are reported in his two-part article "Alphamagic Squares" in *Abacus* (Vol. 4, 1986, pp. 28–45, and 1987, pp. 20–29, 43).

Now for three easy puzzles to work on before the answers are given in the next issue.

1. Construct a 3×3 magic square with nine consecutive positive integers that has a magic constant of 666—the notorious number of the Beast in the Bible's Apocalypse.

2. Place 1 in a corner cell of the 3×3 matrix, then fill the remaining cells with consecutive nonnegative digits to form a magic square.

3. Arrange nine playing cards like this:

1	2	3
3	1	2
2	3	1

All rows and columns have a sum of 6, as well as one main diagonal, but the other diagonal has a sum of 3. Change the positions of three cards to make the square completely magic.

Answers

1.

219	224	223
226	222	218
221	220	225

2. This is the *lo shu* with 1 taken from each number.

1	8	3
6	4	2
5	0	7

3. There are two solutions. Shift the entire bottom row of cards to the top of the square, or move the entire leftmost column of cards to the square's right side.

Addendum

The following addendum to "3 × 3 Magic Squares" appeared in *Quantum* (March/April 1996).

This will update my offer of $100 in "The Magic of 3 × 3" for an order-3 magic square made with nine distinct square numbers. Lee Sallows, mentioned in my article, wrote a program that found many almost magic squares in which only one diagonal failed to give the magic sum. His square with the lowest constant is shown in Figure 1.

Such semimagic squares exist, as John Robertson of Berwyn, Pennsylvania, has shown, if and only if they consist of three triplets of numbers in arithmetic progression, all with the same differences between adjacent terms. Corresponding terms in the triplets need not be in arithmetic progression, as required for the square to be fully magic. Robertson has also shown that finding all such squares is equivalent to finding all the rational points on certain elliptic curves.

127^2	46^2	58^2
2^2	113^2	94^2
74^2	82^2	97^2

Figure 1. *constant* = 147^2

In most cases found by Sallows, the constant is also a square, as in the example given (Figure 1). However, this is not true of all partial magic squares, as shown by the counterexample in Figure 2, discovered by Michael Schweitzer, a Göttingen mathematician.

35^2	3495^2	2958^2
3642^2	2125^2	1785^2
2775^2	2058^2	3005^2

Figure 2. *constant* = 20966014

The constant for rows, columns, and one diagonal is the nonsquare 20966014. In my article I said that order-3 squares made of squares are possible with zero in one cell. I should have added that squares of this type are magic only in rows and columns.

Robertson sent a variety of 4×4 magic squares made with distinct squares, and called my attention to R. D. Carmichael's *Diophantine Analysis*. Order-3 magic squares for powers of n are impossible unless three powers can be in arithmetic progression. For this to be true, the equation $a^n + b^n = 2c^n$ must have solutions with distinct integers for a, b, and c. Leonhard Euler proved there are no solutions for $n = 3$. This rules out order-3 squares made with cubes or multiples of cubes. Carmichael also shows impossibility for $n = 4$ and multiples of fourth powers. I have been informed by Noam Elkies that with Andrew Wiles's proof of Fermat's Last Theorem it can be shown that $a^n + b^n = 2c^n$ has no solution for n greater than 2.

Even though three squares can be in arithmetic progression, there may be no way to construct a 3×3 fully magic square with nine distinct squares. Schweitzer has shown that if such a square exists, the central term must have at least nine digits, and if the entries have no common divisor greater than one, all entries must be odd.

Postscript

By 1900 it was known that no order-3 magic square could be made with powers of 3, 4, and 5. It is now known that the equation $x^n + y^n = 2z^n$ has no solution if n exceeds 2, hence no 3×3 magic square, with distinct digits, can be made with powers greater than 2. The final proof was given by Henri Darmon and Loïe Morol in their paper (it's in English) "Winding Quotients and Some Variants of Fermat's Last Theorem," in *Journal für die reine und angewandte Mathematik* Vol. 490 (1997), pp. 81–100.

My offer of $100 for a magic square made with squares remains. Here I extend the same offer for a proof of impossibility.

John Robertson discusses the $n = 2$ case, and gives the historical background, in "Magic Squares of Squares," *Mathematics Magazine*, Vol. 69 (October 1996), pp. 289–283.

Magic squares made with squares are possible for squares of order 4 and have been constructed for many higher orders. Robertson conjectures that such squares exist for all orders higher than 3.

Chapter 23
Some New Discoveries about
3 x 3 Magic Squares

One afternoon, while sitting in a car waiting for my wife to finish supermarket shopping, I located a pencil and paper and drew on the sheet the unique *lo shu*, or magic square made with distinct digits 1 through 9:

2	7	6
9	5	1
4	3	8

Having nothing better to do, I decided to see what happens if instead of adding digits in each row and column I multiplied them. The top row's product $(2 \times 7 \times 6)$ is 84. The second row's product is 45, and the third row's product is 96. I added the three products to see if the sum was of any interest. It was 225, the square of the *lo shu*'s magic constant of 15. Would the sum of the column products be equally interesting? To my amazement, it also was 225. This equality obviously holds regardless of how the square is rotated or reflected. Could this be just a coincidence? What about the two main diagonals? Might their products also add to 225? No, the sum turned out to be 200.

Back home I began experimenting with other 3 × 3 magic squares having numbers not necessarily in sequence. In every case the sum of the products of rows equaled the sum of the products of columns.

Like so many discoveries in mathematics, far more significant than this, I had found a result by experimenting with numbers in a manner exactly like the way physical laws are discovered. The big difference, of course, is that mathematical conjectures, obtained experimentally, can be confirmed or falsified by rigorous proofs.

This article first appeared in *Math Horizons* (February 1998).

I recalled the following algebraic structure for all order-3 magic squares:

$a+b$	$a-b-c$	$a+c$
$a-b+c$	a	$a+b-c$
$a-c$	$a+b+c$	$a-b$

When distinct real values (including fractions, negative numbers, irrational roots, π, e, and so on), even complex numbers, are substituted for a, b, and c, the result is always magic. The matrix provides a simple though tedious way to prove the conjecture that for all 3×3 magic squares the sum of low products must equal the sum of column products.

I thought I had stumbled on a new property of order-3 magic squares not noticed before, but Lee Sallows, a computer scientist in Holland, quickly disabused me. Hwa Suk Hahn, of West Georgia College, in Carrollton, Georgia, reported the result in "Another Property of Magic Squares" (*The Fibonacci Quarterly*, Vol. 73, 1975, pages 205–208). He called magic squares with the sum of row products equal to the sum of column products a "balanced square," and proved that all order-3 magic squares are balanced. The property was rediscovered by D. B. Eperson, who mentioned it in a brief note to *The Mathematical Gazette* (Vol. 79, July 1995, pages 182–83).

Hahn found similar squares of orders 4 and 5, but not of any higher order. Shown below are balanced squares of orders 4 and 5.

1	14	7	12
15	4	9	6
10	5	16	3
8	11	2	13

1	7	13	19	25
14	20	21	2	8
22	3	9	15	16
10	11	17	23	4
18	24	5	6	12

After reading a first draft of this article, John Robertson suggested adding here the following paragraph:

We have seen above that all order-3 magic squares are balanced while for any higher order, some squares are balanced and some are not. Hahn also found a more subtle difference between magic squares of order 4 and magic squares of higher order. He proved that if a constant is added to every element of any balanced order-4 magic square, the result is always another balanced magic square. If a constant is added to a balanced magic square of order greater than 4 the result will be a magic square, but it might or might not be balanced. If a constant is added to every element of the order-5 magic square in the diagram above, the result is always a balanced magic square. (For proof, note that you need only test five nonzero constants as the sum of the row or column products is a fifth-degree polynomial in the constant.) There is a known order-5 balanced magic square, with non-integral entries, that loses the property of being balanced when a certain constant is added. It would be of interest to find an order-5 balanced magic square with integral entries that produces a non-balanced magic square when some constant is added, or prove there are none.

A question arises. Are there order-3 magic squares for which the sum of the products of the two diagonals is also the same as the other two sums? I was unable to construct such a square.

In correspondence with Robertson, of Berwyn, PA, who is far more skilled than I in solving problems in number theory, I mentioned my discovery about the rows and columns, and wondered if diagonals could have the same magic sum. A few days later he sent a proof that there is an infinity of such squares.

Here's how to construct them. Start with any sequence of three square numbers in arithmetic progression, x^2, y^2, and z^2. For a in the algebraic matrix substitute the value $2y$. For b substitute the value of x, and for c substitute the value of z.

Let's take the simplest example: 1^2, 5^2, 7^2. The three squares, 1, 25, 49, are in arithmetic progression with a difference of 24. Within the matrix, then, a has a value of $2 \times 5 = 10$, b has a value of 1, and c has a value of 7. Lo and behold, like sorcery, the result is a magic square of the type we are seeking (shown at the top of the following page). The sum of the row products is 1,500. This is also the sum of the column products, and the sum of the diagonal products.

Figure 1 shows examples, provided by Robertson, of three higher squares with the same strange property. Above each square I show the

11	2	17
16	10	4
3	18	9

arithmetic progression of square numbers that generates the square, and the values taken by a, b, c in the matrix. Below each square is its magic product–addition sum.

These magic squares—let's call them Robertson squares—are remarkable patterns. Not only are they magic in the traditional sense, they are also magic in an entirely different sense. As known to the ancients, the additive constant of any order-3 magic square is $3a$, or three times the square's central number. Robertson squares have, in addition, what we can call the multiplicative–additive constant. It is $a(2a^2 - b^2 - c^2)$, or more simply, $3a^2/2$.

Here is a simplified account of how Robertson discovered the procedure for constructing such squares. From the algebraic matrix it is easy to determine that the sum of the low (or column) products is $a(3a^2 - 3b^2 - 3c^2)$. As mentioned above, the sum of the diagonal products is $a(2a^2 - b^2 - c^2)$. Setting these equal and simplifying yields $a^2 = 2(b^2 + c^2)$. Because a must be even, we can rewrite the equation as $2(a/2)^2 = b^2 + c^2$.

The last expression tells us that b^2, $(a/2)^2$, and c^2 are squares in arithmetical progression. There is an infinity of such triples, and well-known formulas for producing them. This makes it easy to construct an infinity of Robertson squares.

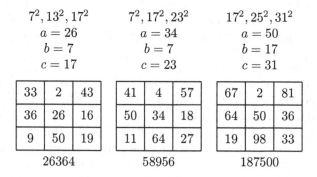

$$7^2, 13^2, 17^2 \qquad 7^2, 17^2, 23^2 \qquad 17^2, 25^2, 31^2$$
$$a = 26 \qquad\qquad a = 34 \qquad\qquad a = 50$$
$$b = 7 \qquad\qquad b = 7 \qquad\qquad b = 17$$
$$c = 17 \qquad\qquad c = 23 \qquad\qquad c = 31$$

33	2	43
36	26	16
9	50	19

26364

41	4	57
50	34	18
11	64	27

58956

67	2	81
64	50	36
19	98	33

187500

Figure 1. Three Robertson Squares.

At the most, three squares can be in arithmetic progression. Such triples are closely related to Pythagorean triangles—right triangles with integral sides. The smallest square root of a number in the progression is the difference between the triangle's legs, the largest square root is the sum of the legs, and the middle square root is the Pythagorean triangle's hypotenuse.

Consider, for example, the familiar 3, 4, 5 Pythagorean triangle. The difference between its legs is 1, the hypotenuse is 5, and the sum of its legs is 7. The squares of these numbers, 1, 25, and 49, form the progression that generates the smallest Robertson square. The three squares shown in Figure 1 correspond to Pythagorean triangles with sides 5, 12, 13; 8, 15, 17; and 7, 24, 25. Thus from any given Pythagorean triangle you can construct a Robertson magic square.

A question remains. Are there order-3 squares other than the *lo shu* such that the sum of the row products equals the sum of the column products, and such that this sum also equals the square of the magic constant? (Diagonals are not considered.) Robertson has shown that the *lo shu* is the only 3×3 square in distinct positive integers that has this property. If zero is allowed, the square below meets the provisos.

14	0	10
4	8	12
6	16	2

The magic constant is 24. The sum of the rows (or columns) is 576, and this equals 24^2. Similar squares exist if negative numbers are allowed. For example, here is such a square found by Sallows.

1	−4	3
2	0	−2
−3	4	−1

Its sum of row products, sum of column products, and sum of diagonal products in each case is zero. Zero also is the square of the magic constant. As Sallows pointed out, adding 5 to each number produces the *lo shu*.

Robertson has found a simple way to construct order-3 magic squares with distinct positive integers such that the sum of the row (or column) products is an integral multiple of the square of the magic constant. He is writing a note explaining the procedure.

An interesting property of order-3 magic squares, long known, is that the sum of the squares of the first row equals the sum of the squares of the third row. This is also true of course of the sums of the squares of the first and third columns, though the two sums are never the same.

Another curious and little-known property of all order-3 magic squares, called to my attention by Monte Zerger, involves treating such squares as matrices. If such a square matrix is multiplied by itself the result is never another magic square. However, if the matrix is cubed, the result is always another magic square! For example, when the *lo shu* is multiplied twice by itself, the result is the following magic square with a constant of $15^3 = 3375$.

1053	1173	1149
1221	1125	1029
1101	1077	1197

For a proof of this theorem as well as other results based on magic matrices see N. Gauthier's paper "Singular Matrices Applied to 3 × 3 Magic Squares," in *The Mathematical Gazette* (Vol 81, July 1997, pages 225–220).

Robertson adds:

For order-3 magic squares that are nonsingular when considered as matrices, it is known that all odd matrix powers are magic squares, including negative powers (Problem E3440, *American Mathematical Monthly*, proposed Vol. 98, No. 5, May 1991, page 437, solved Vol. 99, No. 10, December 1992, pages 966–967). Robertson has shown that an arbitrary matrix product of an odd number of order-3 magic squares is a magic square. It is straightforward to show this for a product of three squares, and then proceed by induction. He conjectures that both of the above results (inverse is magic, arbitrary products of odd numbers of squares are magic) are true of all magic squares of odd order. A proof or disproof would be of interest.

Postscript

Owen O'Shea, a correspondent in Ireland, sent the following curiosities involving the *lo shu*.

For each digit in the *lo shu*, substitute its corresponding Fibonacci number—that is, is position number in the familiar sequence 1, 1, 2, 3, 5, 8, 13, 21, 35.

8	1	21
13	5	2
1	34	3

The sum of the sums of each row is 88, and this is also the sum of the sums of each column. The sum of the products of each row is 600. This is also the sum of the products of each column. The sum of the products of each diagonal is 225, the square of 15.

Cross out the middle column of the *lo shu*, in any of its orientations, to form three two-digit numbers 68, 73, 24.

6	1	8
7	5	3
2	9	4

The sum of 68, 73, and 24 is 165. Reverse each number. The sum of 86, 37, and 42 is also 165.

Consider the following chart of products and sums of the same set of numbers and their reversals:

68^2	$=$	4624		86^2	$=$	7396
73^2	$=$	5329		37^2	$=$	1369
24^2	$=$	576		42^2	$=$	1764
		10529				10529

All these results hold regardless of how the *lo shu* is rotated and reflected.

Chapter 24

Primes in Arithmetic Progression

Euclid proved that the number of primes is infinite, and for more than 200 years it has been known that there are arithmetic progressions that contain an infinite number of primes. A much deeper question, far from resolved, is whether there are arbitrarily long sequences of primes alone that are in arithmetic progression.

In the early 1940s Paul Erdös conjectured that any infinite sequence of increasing integers, such that the sum of their reciprocals diverges, contains arbitrarily long APs (arithmetic progressions). Because the sum of the reciprocals of the primes is infinite (this was proved by Leonhard Euler in 1737) it follows that if Erdös's conjecture is true, the sequence of primes will contain arbitrarily long APs. Although Erdös offered $3000 in 1977 for a proof of his conjecture, the prize has not been collected.

In 1976 L. J. Lander and T. R. Parkin made a stronger conjecture: For every k there are k *consecutive* primes in AP. "This conjecture is undoubtedly true," Erdös has written, "but is completely unattackable by the methods at our disposal." As far as I know, the longest known sequence of consecutive primes in AP are the six terms that start with 121174811, and have a common difference of 30. This was reported by Lander and Parkin in the same paper in which they made their conjecture.

Let k be the number of primes in AP and d be the common difference. If $d = 1$ there is only one sequence: the doublet 2, 3. If $d = 2$ there may or may not be an infinity of what are called "twin primes", or primes that differ by 2. This is the notorious unsolved twin-prime problem. Curiously, the sum of the reciprocals of all twin

This article first appeared in The Rome Press, Raleigh, NC, *Mathematical Sciences Calendar* 1998.

primes *converges*, which suggests (but certainly doesn't prove) that the set of twin primes could be finite.

When $d = 2$ and $k = 3$, the only triplet in AP is 3, 5, 7. With d greater than 2, Sarvadaman Chowla showed in 1944 that the number of prime triplets in AP is infinite. It has long been known that any prime AP starting with 3 cannot have more than three terms, but is there an infinity of such triplets starting with 3? This question remains undecided.

In a note on "Prime Arithmetic Progressions" (*Crux Mathematicorum*, Vol. 7 [1981], p. 68–69) Charles Trigg listed the first 25 triplets that start with 3 in increasing value of the second term.

3	5	7	3	23	43	3	53	103	3	101	199	3	157	311
3	7	11	3	31	59	3	67	131	3	107	211	3	167	331
3	11	19	3	37	71	3	71	139	3	113	223	3	181	359
3	13	23	3	41	79	3	83	163	3	127	251	3	191	379
3	17	31	3	43	83	3	97	191	3	137	271	3	193	383

Prime triples in arithmetic progression starting with 3.

Trigg cited the triplet 3, 5003261, 10006519 as an example of one with a large second term. Surely triplets with larger second terms have since been calculated, but I do not know the current record.

Except for the doublet 2, 3, the difference in any prime AP obviously must be even because otherwise it would produce an even number following a prime. But we can say much more about the difference. It has long been known that if k, the number of terms in a prime AP, is greater than 2, and the difference does not start with k as the first term, d will be divisible by all primes less than or equal to k. If the sequence starts with k, d will be divisible by all primes less than k. This of course sets lower bounds for the difference in any AP of k primes. If the difference is not the minimum set by the product of primes, it will be a multiple of the minimum. For example, suppose k is 25. The difference cannot be less than the product of all primes from 2 through 23, or not less than 223,092,870. If d is more than that, it will be a multiple of 223,092,870. Knowing these divisibility properties greatly simplifies computer searching for long sequences of primes in AP.

Given a sequence of length k, we can ask what sequence of that length maximizes or minimizes certain values such as the common difference, or the first or last term, or the sum of the terms. Here we will consider prime APs with the smallest known first term, given length k. If there is more than one sequence of length k that starts with the same term, we will choose the progression with the lowest difference.

When $k = 3$, the first sequence is 3, 5, 7 ($d = 2$). When $k = 4$ or 5, the first is 5, 11, 17, 23, 29 ($d = 6$). When $k = 6$, the first sequence is 7, 37, 67, 97, 127, 157 ($d = 30$). When $k = 7$, the first is 7, 157, 307, 457, 607, 757, 907 ($d = 150$).

When $k = 8$, 9, or 10, the first sequence starts with 199 ($d = 210$). This progression was first reported in 1910 by E. B. Escott. Magic square buffs realized at once that it made possible a third-order (3×3) square of nine primes in AP. No third-order square of this type has a lesser constant (magic sum of rows, columns, and main diagonals). We shall indicate the constant with c.

1669	199	1249
619	1039	1459
829	1879	409

$c = 3117$

For $k = 11$, the first known prime AP starts with 10,859 ($d = 210$). It has been credited to Malcolm H. Tallman, but I do not know the date of its discovery.

For $k = 12$, the first known starts with 23,143 ($d = 30{,}030$) found by V. A. Golubev in 1967. A 12-term sequence starting with 166,601 has both a lower difference, 11,550, and a lower last term.

When $k = 13$, the first known starts with 4943 ($d = 60{,}060$). I do not know the discoverer.

When $k = 14$, the first known starts with 12,107,737 ($d = 35{,}735{,}700$). It was reported by Sol Weintraub in 1976.

When $k = 15$ or 16, the first known (also discovered by Weintraub and reported in 1976) starts with 13,816,843 ($d = 35{,}735{,}700$). This makes possible the construction of a fourth-order prime square that is also pandiagonal (see below). A pandiagonal magic square (sometimes called a diabolical or Nasik square) is one on which all broken diagonals, as well as the two main diagonals, also add to the magic sum. Put another way, if such a square is drawn on a torus surface, all orthogonal and all diagonal rows add to the same number.

371,173,843	49,552,543	228,231,043	478,380,943
156,759,643	549,852,343	299,702,443	121,023,943
335,438,143	85,288,243	192,495,343	514,116,643
263,966,743	442,645,243	406,909,543	13,816,843

A pandiagonal magic square of primes in arithmetic progression $c = 1,127,338,372$.

It is not hard to prove, from the algebraic structure of third-order magic squares, that no pandiagonal square of this size is possible. The order-4 square, therefore, is the smallest that can be pandiagonal. Whether such a square can be made with a lower constant than one based on Weintraub's sequence of 16 primes in AP remains an open question.

When $k = 17$, the first known prime AP (reported in 1977 by Weintraub starts with 3,430,751,869 ($d = 87,297,210$).

When $k = 18$ or 19, the first known sequence (Weintraub, 1977) starts with 8,297,644,387 ($d = 4,180,566,390$).

When $k = 20$, the first known sequence (found in 1987 by James Fry) starts with 803,467,381,001 ($d = 2,007,835,830$).

If a sequence with k greater than 20 has been found, it has not come to my attention. Of course, if and when a sequence of length 25 is discovered, it will make possible a pandiagonal fifth-order magic square of primes in AP.

Postscript

When I wrote this piece I gave six consecutive primes in arithmetic progression. In 1995 Harvey Dubner and Harry L. Nelson raised the record to seven consecutive primes. The first prime has 97 digits and the difference is 210. In 1997 Dubner and four associates found a sequence of eight consecutive primes, and in 1998 extended the record to nine and ten primes! Along the way their computer found 27 new sequences of eight, and hundreds of sequences of seven.

The jump to eleven consecutive primes is a huge one. Nelson estimates that if such a sequence exists the numbers will be at least a thousand digits long. As Richard Guy wrote in "What's Left?" (*Math Horizons*, April 1998), "It will be a while before anyone finds eleven."

As far as I know the record for non-consecutive primes in arithmetic progression is 22, found in 1993. Guy, in the article cited above, gives two examples. It is still not known if there is a limit on the number of consecutive primes in arithmetic progression.

Chapter 25
Prime Magic Squares

Constructing magic squares with primes not necessarily in AP (arithmetic progression) is one of the more esoteric aspects of magic square theory, though one in which research with the aid of computers can surely make advances that were not within the reach of earlier researchers. Literature on the topic is scattered in obscure places and many questions remain unanswered. Here I will give only a few main results. The famous British puzzle maker Henry Ernest Dudeney, in his *Amusements in Mathematics* (1917), claims that he was the first to consider prime squares (in *The Weekly Dispatch*, July 22 and August 5, 1900). He summarizes some findings, but at that time 1 was considered a prime so many of these early results concern squares in with 1 appears.

Accepting today's convention that 1 is not a prime, the unique third-order prime square with the lowest constant is not given by Dudeney. It is

17	113	47
89	59	29
71	5	101

$c=177$

There are four essentially different forms of the fourth-order prime square with lowest constant. All four are given by Allan W. Johnson, Jr., in *Crux Mathematicorum*, Vol. 5 (1979), page 241. One of them is at the top of the following page.

This article first appeared in *The Mathematical Sciences Calendar* in 1988, issued by the Rome Press (Raleigh, NC).

3	61	19	37
43	31	5	41
7	11	73	29
67	17	23	13

$c=120$

The constant must be doubled to obtain a pandiagonal prime square of order 4. The one shown below was published by Johnson in the *Journal of Recreational Mathematics*, Vol. 12 (1979–80), page 207, and later proved by him to have the lowest possible magic sum for such a square.

41	109	31	59
37	53	47	103
89	61	79	11
73	17	83	67

$c=240$

A second pandiagonal prime square of order 4, with different numbers but the same constant, appears in what I believe to be the first paper ever published on the topic, "Pandiagonal Prime Number Magic Squares" by Charles D. Shuldham, in *The Monist*, Vol. 24 (1914), pages 608–613. (Shuldham's address is given as Wyoming, N.J. Either there was such a town in New Jersey, or it was a misprint for N.Y.) The square is

73	41	13	113
23	103	83	31
107	7	47	79
37	89	97	17

$c=240$

Johnson has proved that the fifth-order prime square shown below has the lowest constant. (It was first published in *Crux Mathematicorum*, Vol. 8 [1982], page 97, in Johnson's article "Magic Squares in Minimal Primes.") Note that its constant also is prime.

3	83	41	101	5
89	67	11	29	37
19	7	103	31	73
79	53	17	13	71
43	23	61	59	47

$c=233$

In the same article Johnson gives what he has proved to be a sixth-order prime square of lowest constant ($c = 432$). A fifth-order pandiagonal square of known lowest constant ($c = 395$) was published by Johnson in the same journal, Vol. 6 (1980), p. 175. His best result for a pandiagonal prime square of order 6 ($c = 486$) has not been published, nor has it been shown minimal.

Little is known about pandiagonal prime squares higher than order 6. Shuldham's examples for orders 7 and 8 use 1 as a prime. He was unable to construct an order 9 with or without 1, but he did construct an order 10, with 7 its smallest prime and a constant of 990.

Even less is known about magic squares made with *consecutive* primes. Johnson has constructed five such squares of order 4, of which the one with lowest constant (published in the *Journal of Recreational Mathematics*, Vol. 14 (1981–82), page 152) is shown below:

37	83	97	41
53	61	71	73
89	67	59	43
79	47	31	101

$c=258$

In his *Crux Mathematicorum* article of 1982, Johnson gave consecutive prime squares for orders 5 and 6, both of which he has shown to have the lowest constants (313 and 484 respectively). Note that the fifth-order square's constant is a prime. *Researches in Magic Squares*, a Japanese book by Akira Hirayama and Gakuho Abe (published in Osaka in 1983), contains different consecutive prime squares for orders 5 and 6 (with the same constants as Johnson's), and consecutive prime squares for orders 7, 8, and 9, with constants of 797, 2016, and 2211 respectively. I do not know whether the last three have been proven minimal.

Is it possible to construct a magic square with consecutive odd primes that start with 3? The surprising answer is yes, but not until a square reaches the size of order 35. Such a square was first constructed in 1957 by Akio Suzuki, of Japan. Another order-35 square of the same sort, but with different numbers, was constructed in 1980 by Captain William H. Benson, a retired U. S. Navy Officer who coauthored, with Oswald Jacoby, two original Dover paperbacks: *New Recreations with Magic Squares* (1976) and *Magic Cubes* (1981). Benson, who has not yet published his square, did not know of Suzuki's earlier discovery.

Does a consecutive-prime magic square exist for order 3? It probably does, but the constraints on the order 3 squares, are so tight that it could be that no such square is possible. It would require three sets of triplets, each triplet in arithmetic progression with the same common difference for each progression, and with the lowest terms of the triplets in an arithmetic progression. If such a square exists, the primes required may or may not be beyond the reach of computer programs that run in a reasonable time. It is an outstanding unsolved problem in magic square theory.

I will give $100 to the first person who can produce an order-3 magic square formed with nine distinct primes that are consecutive.

Postscript

Harry Nelson, in 1988, using a Cray computer at the University of California, won my hundred-dollar prize. He found 22 solutions. What is almost certainly the one with the lowest constant possible for such a square is shown on the following page.

1,480,028,201	1,480,028,129	1,480,028,183
1,480,028,153	1,480,028,171	1,480,028,189
1,480,028,159	1,480,028,213	1,480,028,141

Chapter 26
The Dominono Game

Dominono is a recently discovered relative of tic-tac-toe. Although it appears easy to analyze, it is actually quite complex.

Dominono is played on a square grid such as the two boards shown below. It can be played as a pencil–paper game like tic-tac-toe. Players take turns putting their mark on any vacant square, with one player using X's, the other O's. The person who forms a domino—that is, who marks two squares that share an edge—loses. Note that it's all right to play in squares that only touch diagonally.

It's more fun to play Dominono with checkers, counters, or coins of two different colors. Players take turns placing one of their pieces on a vacant square. The winner is the first to force his opponent to make a domino.

On a 2 × 2 grid, the second player obviously wins. Indeed, if the board's side has an even number of squares, the second player can always win by using a symmetry strategy. After each move by his opponent, he simply plays on the square symmetrically opposite the opponent's move on a line drawn through the board's center. Because of this strategy, only odd-sided boards offer playable games.

On a 3 × 3 board, the second player wins when both sides play their best, but it takes considerable analysis to prove it.

If the opening move is in the center, the second player wins by taking, on his or her first two turns, two corner squares that are not diagonally opposite one another. Playing a third move on any safe square then forces the opponent to make a domino.

If the opening move is on a side square, the first player clearly cannot later take the center without losing. The second player, therefore, wins by playing symmetrically.

If the opening move is in a corner, I thought there was no simple winning strategy, until Fred Galvin, a mathematician at the Univer-

---

This article first appeared in *Games Magazine* (April 1999, p. 53).

sity of Kansas, surprised me by finding a succinct strategy that takes care of all opening moves. Here are the rules, based on the square coloring shown below.

1. Play on a white square as long as possible.
2. Never let your opponent take two opposite corners.
3. Don't make a domino.

After you run out of white squares, if your opponent hasn't already lost, he or she will lose on his next move.

The 5 × 5 board shown, as well as all larger boards, remain unsolved. Perhaps a reader can write a computer program that will decide whether the first or second player can always win, or whether the game is a draw if both sides play perfectly. It has been conjectured that the second player can always win on all square boards, but this is far from established. Try playing on the 5 × 5. You'll quickly see how enjoyable the game is even on so small a board, and also how complex!

⋆   ⋆   ⋆

Now for a puzzle. Suppose the first player takes the square marked X on the 5 × 5 board. Can you prove that this is a losing opening?

⋆   ⋆   ⋆

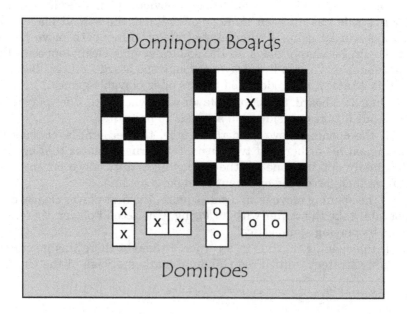

Answer: like the side opening on the 3×3, this starting move prevents the first player from later taking the central square. The second player wins easily by symmetry play.

Postscript

On all even/even rectangular boards there is a central intersection point that allows the second player to force a win by always playing symmetrically opposite that point after each of his opponent's moves. If the rectangle is odd/odd, or odd/even, no general strategy is known, and only small fields have been solved.

I explained how the second player wins on the 3×3. A computer program by A. E. Brouwer, in Holland, proved a second player win on the 5×5. Fred Galvin, in Lawrence, Kansas, showed by hand that the second player also wins on the 3×5 and 3×7, but the 5×7 and 7×7 remain unsolved. He conjectures that the second player wins on all odd/odd boards.

On even/odd boards the second player has an easy win on the 2×3. Galvin found second player wins on the 2×5, 3×4, 3×6, and 4×5. The 4×7 case is still open. Galvin has proved that on all even/odd boards the second player can either win or draw, but never lose.

On $1 \times n$ fields, Galvin confirmed an earlier result by Richard Guy that the game is a draw if $n = 1, 2, 4, 6$. Otherwise the second player wins.

It is surprising that a game with such simple rules is so complicated and intractable on odd/odd and odd/even boards.

Chapter 27
The Growth of Recreational Mathematics

Amusement is one of the fields of
applied mathematics,
—William F. White

My "Mathematical Games" column began in *Scientific American's* December 1956 issue with an article on hexaflexagons, curious hexagonal paper structures that keep changing their faces when properly flexed. Physicist Richard Feynman was one of their inventors. At that time only a few books on recreational mathematics were in print. W. R. Rouse Ball's classic *Mathematical Recreations and Essays* (1892) had been updated by the Canadian geometer H.S.M. Coxeter. A translation from the French of Maurice Kraitchik's *Recreational Mathematics* was available as a Dover paperback. Dover had not yet reprinted Henry Dudeney's *Amusements in Mathematics*, or my two-volume selection of mathematical puzzles by Sam Loyd. Aside from a few other puzzle collections, that was about it.

Since 1956 there has been a remarkable explosion of books about recreational math. Many are by such distinguished mathematicians as Ian Stewart, who for some time wrote *Scientific American's* column on the topic; Clifford Pickover, of IBM's Laboratory; John Conway, of Princeton University; Richard Guy, of Calgary University; and Elwyn Berlekamp, of the University of California at Berkeley. *Mathematics: Problem Solving Through Recreational Mathematics*, a pioneering textbook by Bonnie Averbach and Orin Chein, of Temple University, was published by W. H. Freeman in 1980. New books in the field are being written every year by mathematicians too numerous to mention.

A heavily revised and cut version of this paper appeared in *Scientific American* (Vol. 279, August 1998).

Articles on recreational topics run with increasing frequency in mathematical periodicals. The quarterly *Journal of Recreational Mathematics* began publishing in 1961. The National Council of Teachers of Mathematics (NCTM) issued the first volume of William Schaff's extensive *Bibliography of Recreational Mathematics* in 1955. New references have required three sequels.

Obviously no sharp line separates entertaining math from serious math. One reason is that creative mathematicians regard their work as much a form of play as golf is to a professional golfer. In general, math is considered recreational if it has a strong aspect of play that can be appreciated by any layman interested in mathematics. This can take many forms. They include elementary problems with elegant, at times surprising, solutions; mind-bending paradoxes, bewildering magic tricks, games that are easy and fun to play, ingeniously concealed fallacies; topological curiosities such as Möbius bands and Klein bottles; puzzles involving placing or moving counters, sliding blocks, mazes, logic, dissections, tiling, and many other categories. Almost every branch of mathematics below calculus has areas that can be considered recreational. Let me give four amusing examples.

⋆ ⋆ ⋆

Jones puts three cards face down in a row. Only one is an ace, but you don't know which. You place a finger on a card. The probability it is on the ace clearly is 1/3. Jones now secretly peeks at each card, then turns face up a card that is *not* the ace. What is the probability your finger is on the ace? Most everyone thinks it has risen from 1/3 to 1/2. After all, only two cards are face down, and one must be the ace. Actually, the probability remains 1/3. If you now switch your finger to the other face-down card, the probability it is on the ace jumps to 2/3.

I introduced this problem in my *Scientific American* column (October 1959) in the form of a warden and three prisoners. In 1990 Marilyn vos Savant, in her popular *Parade* column, gave it with three doors and a car behind one of them. Her answer was correct, but she received thousands of angry letters, many from Ph.D. mathematicians, berating her for her ignorance of elementary probability! The fracas generated a front-page story in the *New York Times*.

⋆ ⋆ ⋆

The statement "All crows are black" is logically equivalent to the statement "All nonblack objects are not crows." Finding a black crow obviously confirms the first statement. You pick up a red tomato and see that it is not a crow. This just as obviously confirms the second statement. Is a red tomato, therefore, a confirming instance of "All crows are black?" If not, why not?

My April 1957 column introduced this question, known as Hempel's paradox after philosopher Carl Hempel, who discovered it. I am still getting letters about it, and a raft of papers claiming to resolve the paradox have appeared in philosophical journals.

★   ★   ★

Arrange a deck so the card colors alternate. Cut the deck about in half, making sure that the bottom cards of each half are not the same color. Riffle shuffle the two halves together as thoroughly or as carelessly as you please. Take cards from the top of the deck in pairs. Each pair will consist of a black and a red card! This is known to magicians as the Gilbreath principle after its discoverer, Norman Gilbreath. I first explained it in my column of August 1960, discussed it again in July 1972, and added a proof when the last column was reprinted in *New Mathematial Diversions*.

Magicians have invented more than a hundred card tricks based on this principle. It generalizes. Arrange the deck so the suits are in cycles of, say, spades, hearts, clubs, and diamonds. Deal any multiple of four cards to the table, thus reversing the suit ordering, until you have a pile of 24 or 28 cards. Riffle shuffle the two piles together. Every quadruplet of cards, taken from the top, will contain the four suits. Try this with two decks. Arrange one deck in any random order, and the other deck with cards in the reverse order. Shuffle the decks together, then divide the cards exactly in half. Each half will be a complete deck!

Although Gilbreath had no purpose in mind other than to furnish a basis for card tricks, last year Donald Knuth, Stanford University's noted computer scientist, discovered a wonderful application of the principle to the efficient input and output of computers.

★   ★   ★

Consider the square patterns shown in Figure 1. Each is made with the same sixteen pieces, yet the square on the right has a large hole in its center! What happened to the missing area? It would spoil the fun to explain it here. This strange fallacy, which I devised for a May 1961 column, was recently made in China as a puzzle with colorful plastic pieces, for sale in a dollar store chain.

★   ★   ★

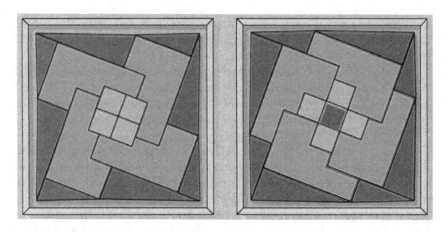

Figure 1. The author's vanishing area paradox. The 16 polygons on the left form a solid square. The same 16 pieces, in a different pattern on the right, form a square with a central hole. Where did the missing area go?

Although *Mathematics Teacher*, a monthly published by the NCTM, has increasingly carried articles on recreational topics, teachers themselves have, by and large, ignored such material. For forty years I have done my best to convince them that entertaining math is more than what they call "enrichment" material. It should be introduced regularly as a way to arouse the interest of young students in the wonders of mathematics. So far, movement in this direction has been at a snail's pace. I have often told the story of how one day, in a high school math study period, I had finished my assignment and was trying to decide whether the first player can always win a game of tic-tac-toe. The teacher saw my scribbling, snatched away the sheet, and said, "Mr. Gardner, when you're in my class I expect you to work on mathematics and nothing else."

The situation is much the same today. I know of no recommendation by the NCTM that asking students to solve the game of tic-tac-toe is a superb way to introduce combinatorial mathematics, symmetry, graph theory, game theory, and even probability. Moreover, it ties in strongly with every student's experience. Who has not, as a child, played tic-tac-toe?

I recently purchased the 1997 yearbook of the NCTM, titled *Multiculturalism and Gender Equity in the Mathematical Classroom: The Gift of Diversity*. The book has 28 chapters written by teachers who defend a current trend that goes by the name of "new new math" to distinguish it from the failures of the old "new math" that had been so disastrously backed by the NCTM in the late 1960s. The new fad tries to do away with lecturing by forming the class into small groups of students who are given problems and asked to solve them by cooperative reasoning.

For example, instead of explaining the Pythagorean theorem on a blackboard, students are asked to cut right triangles and squares from graph paper and discover the theorem all by themselves. This may take days, and even end in failure. "Interactive learning," as it is called, is substituted for passive listening to a teacher. Rote memory of addition and multiplication tables, and learning how to do long division, is replaced by the use of calculators. The result: When students get to college they are unable to multiply or add even small numbers without reaching for their calculator. As for long division, forget it!

The new new math also places a heavy emphasis, admirably so, on multiculturalism and gender equity. Translated from the yearbook's mind-numbing jargon, this means that all ethnic groups should be treated equally, and girls treated the same as boys. It is the new new math's positive side.

What struck me most about the yearbook was that not a single teacher had anything good to say about the value of recreational math in hooking student interest. Let me propose to teachers the following experiment. Ask each group of students to think of any three-digit number ABC, then put into their calculator the number ABCABC. Thus if they thought of 237, they would give their computer the number 237237. Tell them you have the psychic power to predict that if they divide their ABCABC number by 13 there will he no remainder. This proves to be true. Now ask them to divide the result by 11. Again, as you predicted, no remainder. Finally, ask them to divide by 7. Lo and behold, their original ABC number is now in the readout. I'll wager that this task will create much more excitement than asking them to invent the Pythagorean theorem. I know of no better introduction to number theory and the properties of primes than asking students to explain why this trick always works.

One of the greatest joys of writing the *Scientific American* column over a period of thirty years was getting to know so many authentic mathematicians. I myself am little more than a journalist who loves mathematics and can write about it glibly. I took no math in college. If you go through my fifteen book collections of columns you will see that the math grew increasingly sophisticated as I learned more. But the key to the column's popularity springs not so much from my writing as from the material I was able to coax from genuine mathematicians.

Solomon Golomb, of the University of California, was one of the first to supply grist for the column. I had the pleasure of introducing his polyominoes, shapes formed by joining unit squares along their edges. Polyomino problems are now a flourishing branch of recreational math. Two books are devoted entirely to these fascinating little shapes, one by Golomb, one by George Martin, at SUNY Albany. Golomb also provided material on what he named "rep-tiles," identical polygons that fit together to form larger replicas of themselves. Five examples are shown in Figure 2. Any rep-tile obviously will tile the plane by making larger and larger replicas.

The late Piet Hein, Denmark's illustrious poet and inventor, became a good friend. I devoted a column to his topological game of Hex (July 1957 and June 1975), to his superellipse and superegg (Sept. 1965), to his Soma Cube (Sept. 1958, July 1969, and Sept. 1972), and other of his inventions. There is a beautiful proof that the first player can always win Hex on a board of any size, though it tells you nothing about how to go about it. The Soma Cube consists of seven polycubes— unit cubes joined at their faces. They fit together like the seven "tans" of tangrams to form a cube, as well as endless other forms. Conway was among the first to prove that the cube has exactly 240 solutions.

John Conway is one of the world's undisputed geniuses. He came to see me one day and asked if I had a Go board. I did. On it he demonstrated his now-famous cellular automata game called Life. By applying three ridiculously simple rules that tell when a cell on the board changes to full or empty, an astonishing variety of Life forms are created, some that move across the field like tiny insects. That column became an instant hit with computer buffs. It is said that for many weeks business firms and research labs were almost shut down while Life enthusiasts experimented with Life forms on their computer screens.

Conway later invented an entirely new way to define numbers that generates infinite classes of weird numbers never met before. I explained this in a September 1976 column. Donald Knuth wrote a novel entitled *Surreal Numbers* about how a boy and girl found some ancient stone tablets that enabled them to develop Conway's number system.

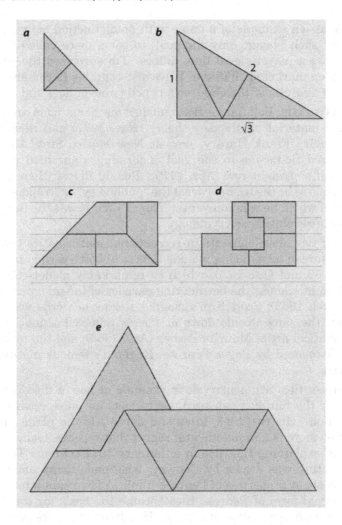

Figure 2. Five low-order rep-tiles. (a) A rep-2 triangle. (b) A rep-3 triangle. (c) A rep-4 quadrilateral. (d) A rep-4 hexagon. (e) The Sphinx: the only known rep-4 pentagon.

Conway collaborated with Guy and Berlekamp on a two-volume work called *Winning Ways* (1982)[1] which I consider to be the greatest contribution to recreational mathematics this century. It is a wondrous source of completely new material. One of its hundreds of gems is a two-person board game called Phutball. Phutball was invented by

---

[1] Reissued by A K Peters, Ltd. in a four volume edition.

Conway as an example of a game with no distinction between moves made by each player; only the goal, to get a piece called the phutball across a player's goal line, differs. I'm surprised the game has never been marketed. *Winning Ways* also contains Berlekamp's pathbreaking analysis of the children's pencil game of dots and squares.

Ron Graham, Bell Labs's distinguished mathematician and juggler, supplied material on Ramsey theory (Nov. 1977) and Steiner trees (May 1986). Frank Harary, now at New Mexico State University, generalized tic-tac-toe to the goal of forming a specified polyomino instead of a straight row (Apr. 1979). Ronald Rivest, then at M.I.T., allowed me to be the first to reveal the "public-key" or "trapdoor cipher" system of which he was coinventor (Aug. 1977). It was the first of such ciphers to revolutionize cryptology.

Other new developments in recreational mathematics that I had the pleasure of introducing include Robert Abbott's card game Eleusis (June 1959 and Oct. 1970) which so remarkably models the role of induction in science; the bewildering paradox of the unexpected hanging (March 1963); Scott Kim's magic lettering of words and phrases that are the same upside down or mirror reflected (June 1981); the mathematical art of Maurits Escher (Apr. 1966); and the nonperiodic tiling discovered by Roger Penrose, England's famous mathematical physicist.

Penrose tiles are a marvelous example of how a discovery made solely for the fun of it can turn out to have an unexpected practical application. His two tiles, kites and darts, tile the plane in such a way that there is no fundamental region that repeats itself. I had the honor of explaining the tiles in a January 1977 column. The issue's cover picture was drawn by Conway, who made many amazing discoveries about properties of Penrose tiling. A few years later, a three-dimensional form of Penrose tiles became the basis for constructing a hitherto unknown type of crystal. Hundreds of papers on what are now called quasicrystals have been published. The outstanding unsolved question in tiling theory is whether a single tile exists that can tile the plane only nonperiodically.

The two columns that generated the greatest number of letters were my hoax column and one on Newcomb's paradox. The April Fool column (Apr. 1975) purported to cover great breakthroughs in science and math. They included a refutation of relativity theory, a psychic motor, the discovery that Leonardo da Vinci had invented the flush toilet, that the opening move of pawn to king rook four was a certain win in chess, and that e raised to the power of $\pi\sqrt{163}$ was exactly equal to the integer 262,537,412,640,768,744. To my amazement, thousands of readers failed to recognize the column as a joke. I gave a picture of a complicated map that I said required five colors. Hundreds of readers sent me copies of the map colored with four colors, many saying the task had taken days.

Newcomb's paradox, named after its discoverer physicist William Newcomb, was first described in a technical paper by Harvard philosopher Robert Nozick. I received so many letters after writing about the paradox (July 1973) that I packed them into a large carton and personally delivered them to Nozick, who analyzed them in a guest column (Mar. 1974).

The paradox concerns two boxes, A and B. Box A contains a thousand dollars. Box B may either contain a million dollars or be empty. You are allowed to choose either B, in the hope of getting a million, or you can take both boxes.

A superbeing, God if you like, has the power of knowing in advance how you will choose. If he predicts that out of greed you will take both boxes, he leaves B empty and you get only the thousand in A. If he predicts you will take only Box B, he has put a million dollars in it. You have watched this game played many times with others. In every case if a player took both boxes, he or she found the box empty. Every time a player took only B, he or she became a millionaire.

How should you choose? The pragmatic argument is that because of the previous games you have witnessed, you can assume that the superbeing does indeed have the power to make accurate predictions. You should, therefore, take only B and know you will get a million. But wait! The superbeing made his prediction *before* you play the game, and has no power to alter it. There are two possibilities. Either B is empty or it contains a million. If empty, you lose nothing by taking both boxes. If it contains a million, you gain a million plus a thousand by taking both boxes.

Each argument seems impeccable. Yet each cannot be the best strategy. Nozick concluded that the paradox, which belongs to a branch of mathematics called decision theory, remains unresolved. My personal opinion is that the paradox proves, by leading to a logical contradiction, the impossibility of a superbeing's ability to predict a person's decision. In any case, the paradox remains controversial, and a raft of papers have grappled with it since *Scientific American* gave it wide currency.

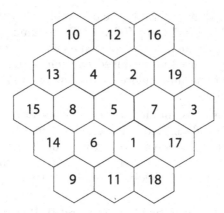

Figure 3. The only magic hexagon possible with consecutive number starting with 1. Every straight row of 3, 4, or 5 cells has a sum of 38.

Magic squares have long been a part of recreational math. The order-3 (3 × 3) square, using digits 1 through 9, is unique apart from rotations and reflections. Squares of order 4 have 880 patterns, and the number of arrangements increases rapidly for higher orders. Suprisingly, this is not the case with magic hexagons. When I published the order-3 hexagon shown in Figure 3 (Aug. 1963), Charles Trigg, a retired Californa math teacher, proved that not only is this elegant pattern the only order-3 magic hexagon, but that no magic hexagons of any other size are possible.

Magic squares need not be restricted to numbers in consecutive order starting with 1. If the only requirement is that the numbers be distinct, a wide variety of order-3 magic squares, in which rows, columns, and the two main diagonals have the same sum, can be constructed. For example, there is an infinity of such squares with distinct prime numbers. Can an order-3 square be made with nine distinct square numbers? A few years ago I offered $100 for such a pattern. It has been proved that no 3 × 3 magic square can consist of distinct cubes or any higher powers, but no one has yet proved impossibility for a "square of squares." If it exists, its numbers will be huge, perhaps beyond the reach of today's supercomputers. Such a square would be as useless as the magic hexagon. Why, then, are mathematicians trying to find it? Because it might be there.

Every year or so I would devote a column to an imaginary interview with a numerologist I called Dr. Irving Joshua Matrix. (Note the 666 provided by the letters of his names.) The doctor would expound on unusual properties of numbers as well as bizarre forms of word play. Many readers thought Dr. Matrix and his beautiful half-Japanese daughter Iva Toshiyori were real. I recall a letter from a puzzled Japanese reader who told me that Toshiyori was a most peculiar surname in Japan. I had taken it from a map of Tokyo. My informant said that in Japanese the word means "street of old men."

I regret I have not asked the doctor his opinion on the recent preposterous bestseller *The Bible Code*. It concerns a cipher system Dr. Matrix could have invented and applied, not just to the Old Testament, but also to the New Testament, the Koran, to great epics of poetry, or even to the absurd book in which it is detailed.

The last time I heard from Dr. Matrix he was in Hong Kong investigating the accidental appearance of π in well-known works of fiction. He cited, for example, the following sentence in Part 2, Chapter 9, of H. G. Wells's *War of the Worlds*: "For a time I stood regarding...." The letters in the words give π to six digits!

Postscript

It would of course require a book to report on high points in the development of recreational mathematics during recent decades. I can only refer readers to the books I mentioned, books I didn't mention, and the sixteen volumes that are collections of my *Scientific American* columns over a period of a quarter-century.

I cannot, however, resist adding a note about my final paragraph. It is hard to believe, but Michael Keith, writing in *Word Ways* (August 1999, pp. 189–90) bettered the H. G. Wells quote by finding two literary sentences in which the word order gives π to seven digits! They are:

"For I have a great sacrifice to...." (King James Bible, I Kings 10:19).

"And a pair O' Hells wherewith to...." (Rudyard Kipling's poem "Ballad of the Clamperdown," line 4).

How Keith located these sentences boggles my mind.

Chapter 28
Maximum Inscribed Squares, Rectangles, and Triangles

It is easy to construct the largest rectangle that will fit inside a given triangle. Simply draw lines from the midpoints of any two sides to meet the third side perpendicularly as shown in Figure 1. Rectangle ABCD will be the largest rectangle that will go inside the triangle. There are three such rectangles (not necessarily the same shape) in any acute triangle. Right triangles have two, and obtuse triangles have only one, which rests on the side opposite the obtuse angle.

It is not hard to show by calculus or algebra that the area of the maximum rectangle is exactly half the triangle's area, and the rectangle's base is half the triangle's base. This result goes back to Euclid's Book 6, Proposition 27. A neat way to demonstrate it is by paper folding. Cut the triangle from paper, then fold over the three corners as shown in Figure 2, folding along the rectangle's three interior sides.

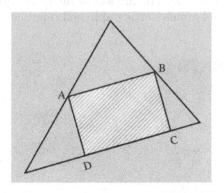

Figure 1. Constructing a maximum rectangle in any triangle.

This paper first appeared in *Math Horizons* (September 1997).

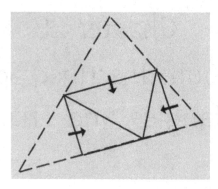

Figure 2. Paper fold demonstration that the maximum rectangle is half the triangle's area.

The flaps will fit snugly into the rectangle. This provides a simple way to construct a maximum rectangle by paper folding.

Paper folding also will demonstrate that no other inscribed rectangle can be larger. Try drawing one that is shorter and fatter, or taller and thinner. Fold over the corners as before by folding along the new rectangle's three sides. You'll find that the flaps either overlap each other, or overlap the rectangle, showing that its area is less than half that of the triangle.

A question now arises. Could there be a larger triangle entirely inside—that is, one that docs not have a side resting on a side of the triangle? (See Figure 3). The answer is no, although proving it is not so easy. You'll find proofs in the published references listed at the end

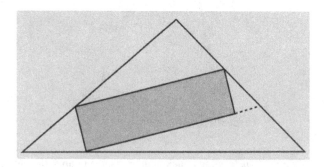

Figure 3. An interior rectangle not inscribed.

of this article. An informal demonstration can be made by paper fold-ing. Cut along the dotted line, then fold over the flaps by folding along the rectangle's sides. You'll find the flaps overlapping the rectangle, proving that their combined area exceeds the rectangle's area.

It is not possible, by the way, for any rectangle to be inscribed in any triangle without resting on one of the triangle's sides. Do you see why? A polygon is said to be "inscribed" in a larger polygon only if all its vertices are on the perimeter of the circumscribed polygon. Apply the pigeonhole principle. A rectangle has four corners but a triangle · has only three sides, therefore two corners must touch the same side.

When we seek the largest square that will go inside any given tri-angle, the task becomes more interesting. Like maximum rectangles, such squares must rest on a side of the triangle. Call this the triangle's base. Let a represent the base and b the triangle's altitude. The for-mula giving the side of the maximum inscribed square is wonderfully simple:

$$\frac{ab}{a+b}$$

The late geometer Leon Bankoff, in a letter cited here as a refer-ence, provided two algebraic proofs of this formula. The simpler one makes use of the triangle shown in Figure 4 with its largest inscribed square. Triangle Ade is similar to triangle ABC, so we can write

$$\frac{b-x}{x} = \frac{b}{a}$$
$$xb = ab - ax$$
$$ax + xb = ab$$
$$x(a+b) = ab$$
$$x = \frac{ab}{a+b}$$

There are different ways to construct such maximum squares. One of the simplest is shown in Figure 5 with respect to acute, right, and obtuse triangles. Erect a square on the outside of the triangle's base, then draw lines from A and B to C. They will intersect the triangle's base at points that mark the base of the side of the largest square resting on that base. (Another way to construct such squares can be found in George Polya's *How To Solve It*, 1945, Part 1, Problem 18).

Unlike maximum rectangles in triangles, maximum squares that are on different sides of a triangle need not have the same area. On

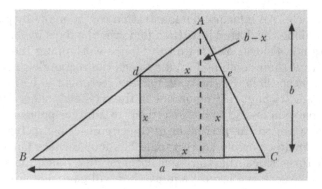

Figure 4. Algebraic proof that $x = \frac{ab}{(b+c)}$.

the right isosceles triangle in Figure 6, for example, the square shown is slightly larger than the largest square resting on the triangle's hypotenuse.

Figure 7 displays the largest square that goes inside an equilateral triangle of side 1. It is slightly smaller than the largest inscribed rectangle. On this triangle the three maximum squares on the triangle's three sides are, of course, equal. Is this the case only for the equilateral triangle?

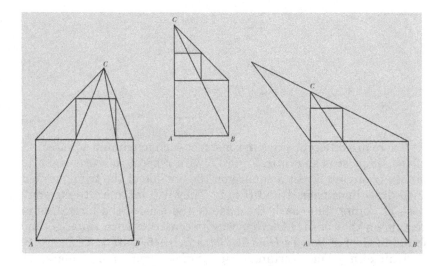

Figure 5. Constructing the maximum interior square on a triangle's side.

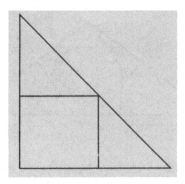

Figure 6. Largest square in a right isosceles triangle.

Amazingly, the answer is no. There is just one other triangle for which this also is true. It is the obtuse isosceles triangle shown in Figure 8. I found it in *The Book of Numbers*, by John Conway and Richard Guy (1996), page 206, where it is credited to Eugenio Calabi, a geometer at the University of Pennsylvania. The ratio of base to side is 1.55138 ..., a number that is the positive value of x in the equation $2x^3 - 2x^2 - 3x + 2 = 0$.

The formula $ab/a + b$, applied to the largest square within the equilateral triangle of side 1, gives the square's side as $\sqrt{12} - 3 = 2\sqrt{3} - 3 = 0.46410161513 \ldots$ Note this number carefully. We will meet it again in a completely unexpected place.

How can one construct a triangle of largest area that will fit inside a given rectangle? Simply draw lines from the corners of the rec-

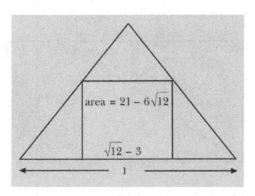

Figure 7. Dimensions of the largest square in an equilateral triangle of side 1.

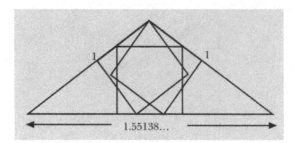

Figure 8. The only triangle, aside from the equilateral, with three equal maximum squares.

tangle's base to any spot on the opposite side. There obviously is an uncountable infinity of such triangles. Like rectangles of the largest area in any triangle, the largest triangles in any rectangle are each half the rectangle's area. We can't demonstrate this by folding, but if you snip off the two small triangles at each side of a maximum triangle, they will fit neatly into the large triangle. Figure 9 shows how this can be proved geometrically. Triangles a and a', and triangles b and b', obviously are congruent.

Consider now the largest equilateral triangle that will fit within a unit square. Slightly larger than such a triangle resting on a square's base, it is shown in Figure 10. To be of largest area its corners must touch all four sides of the square, but since a square has four sides, one of the triangle's corners must be at a corner of the square. An informal pr oof that this triangle cannot be enlarged is to imagine it rotated in either direction, keeping its lower left corner fixed and sliding the

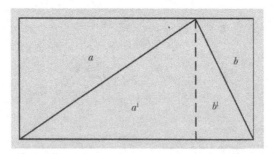

Figure 9. Proof that the maximum-area triangle inscribed in any rectangle has an area equal to half the rectangle's area.

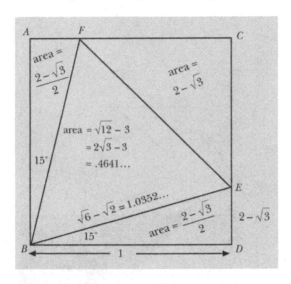

Figure 10. The largest equilateral triangle inside a unit square.

other two corners along the square's other two sides. Obviously this will make one side of the triangle shorter than the other two so the triangle will no longer be equilateral. I do not know if a formal proof by calculus would be easy or hard.

The side of the equilateral triangle is the secant of 15°, which is $\sqrt{6} - \sqrt{2} = 1.03527\ldots$. We can now determine the triangle's area by finding its altitude ($\sqrt{6 - 3\sqrt{3}} = 0.8965\ldots$) then halving the product of altitude and base, or by using Heron's formula for the area: $\sqrt{s(s-a)(s-b)(s-c)}$, where s is the semiperimeter (half the perimeter of the triangle), and a, b, c are the triangle's sides. The area of the maximizing triangle turns out to be $2\sqrt{3} - 3$, the same as the *side* of the largest square that fits into an equilateral triangle of side 1! Has this been noticed before? I found it extremely mystifying until Princeton's John Conway came up with a beautiful proof that I will explain in a moment.

Abul Wefa (A.D. 940–998), a Persian geometer, gave five different ways to construct this largest equilateral triangle with compass and straightedge. Three are reproduced in David Well's *Penguin Book of Curious and Interesting Puzzles* (1992, Puzzle 38). Wells refers readers to *Episodes in the Mathematics of Medieval Islam* (1986), by J.L. Berggren, a book I have not seen.

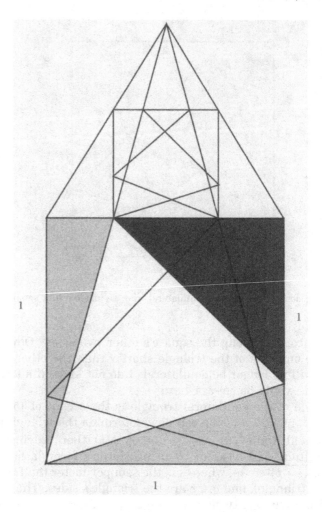

Figure 11. A pattern linking the largest square in an equilateral triangle of side 1 with the maximum equilateral in a unit square.

Henry Ernest Dudeney, in *Puzzles and Curious Problems* (1931, Problem 201) shows how to construct the triangle by folding a square sheet of paper.

An easy way to draw the triangle (I do not know if it is one of Wefa's methods) is shown in Figure 11. As science writer Barry Cipra discovered, the same lines that construct the largest square in the equilateral triangle also mark the points on the top of the larger square

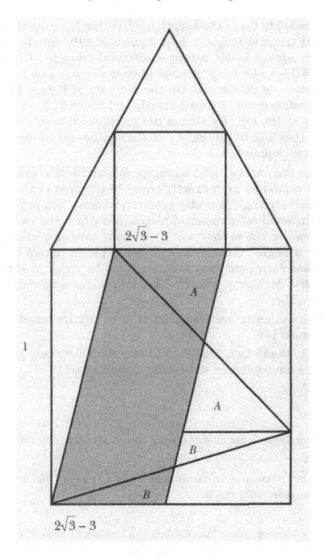

Figure 12. John Conway's dissection proof of a curious equality.

that are the top corners of the two maximum equilateral triangles that fit within the unit square. By drawing these two triangles we get a lovely bilaterally symmetrical pattern that shows how closely related the two constructions are.

I mentioned to Cipra the surprising fact that the area of the largest equilateral triangle inside a unit square exactly equals the side of the largest square inside a unit equilateral triangle. Cipra in turn mentioned this to Conway. Almost at once Conway saw how to prove the equality by a dissection. On the diagram of Figure 12 draw the shaded parallelogram. Its area clearly is 1 times $2\sqrt{3} - 1$, the side of the square at the top. By slicing the parallelogram into three pieces as shown, they will fit precisely into the equilateral triangle inscribed in the bottom square!

Because the narrow-right triangles colored in light gray in Figure 11 have a combined area exactly equal to the area of the dark gray right isosceles triangle, a pretty geometrical dissection problem arises. What is the smallest number of pieces into which the two light gray triangles can be cut so they will fit without gaps or overlaps into the dark gray triangle? Clearly the number of pieces cannot be less than four because each light gray triangle must be sliced at least once to fit inside the dark gray triangle. A solution is possible in four pieces (see Figure 13).

Are there other interesting properties or puzzles based on the pattern of Figure 11?

I wish to thank Don Albers and Peter Renz for many good suggestions that I have followed in writing this article.

References

[1] Bankoff, Leon. Letter in *Mathematics Teacher* 79, May 1986, 322.

[2] Bird, M.T. "Maximum Rectangle Inscribed in a Triangle." *Mathematics Teacher* 24, December 1971, 759–760.

[3] Embry-Wardrop, Mary. "An Old Maxi-Min Problem Revisited." *American Mathematical Monthly* 97, May 1990, 421–423.

[4] Lange, Lester H. "What Is the Biggest Rectangle You Can Put Inside a Given Triangle?" *Mathematics Teacher* 24, May 1993, 237–240.

[5] Niven, Ivan. *Maxima and Minima Without Calculus.* Mathematical Association of America, 1981.

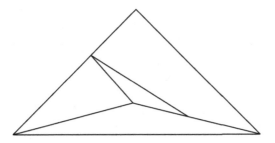

Figure 13. Solution to dissection problem.

Postscript

Solomon Golomb pointed out that it is not strictly accurate to say that the area of the largest equilateral triangle inside a unit square exactly equals the side of the largest square inside a unit equilateral triangle. Reason: One is a linear measurement, the other an area. A more precise way to put it is to think of $2\sqrt{3} - 3$ as two equal ratios.

Tom M. Apostol and Mamikon Mnatsakanian, writing on "Triangles in Rectangles" (*Math Horizons*, February 1998) explained a variety of unusual theorems related to this chapter. For example, my Figure 10 shows a special case of the following theorem: "An equilateral triangle inscribed in, and having a common vertex with a rectangle, cuts off three right triangles inside the rectangle." Call the areas A, B, and C, as shown in Figure 14. Areas $A + B = C$. The article goes on to state other related theorems.

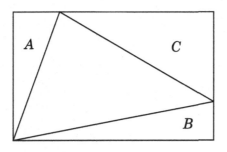

Figure 14. $A + B = C$.

Chapter 29
Serial Isogons of 90 Degrees

In 1988 the first author devised a computer program to search a unit-square grid for closed paths with the following properties. The path starts along a lattice line with a segment of unit length, turns 90 degrees in either direction, continues for 2 units, turns again in either direction, continues for 3 units, and so on. In other words, the segments of the path are in serial order 1, 2, 3, ... N, with a right angle turn at the end of each segment. A path of N segments—the number is of course the same as the number of turns or corners—is said to be a path of order N.

If the path returns to its starting point, making a right-angle with its first segment, we call it a *serial isogon* of 90 degrees. The isogon is allowed to self-intersect, to touch at corners, and to overlap along segments. Think of it as a serial walk through a city of square blocks and returning to the starting corner, or as the moves of a rook on a large enough chessboard.

It is not obvious that such paths exist. However, with a little doodling you will discover the unique isogon of order 8, the lowest order a 90-degree isogon can have. It outlines a polyomino of 52 unit squares that, as Figure 1 shows, tiles the plane. Indeed, it satisfies the "Conway criterion" [1] for identifying tiling shapes. The boundary of the polyomino can be partitioned into six parts, the first and fourth of which ($\overline{AB}$ and $\overline{ED}$ in Figure 2) are equal and parallel, while the other four ($BC = 3$; $CD = 4, 5, 4$; $EF = 7$; $FA = 1, 8, 1$) each have rotational symmetry through 180° about their midpoints (black dots in Figure 2). It may be the only plane-tiling polyomino with a serial boundary, but we are unable to prove this.

Puzzle: Can you tile this polyomino with thirteen L-shaped tetrominoes?

This article, coauthored with Lee Sallows, Richard Guy, and Donald Knuth, first appeared in *Mathematics Magazine* (Vol. 64, December 1991).

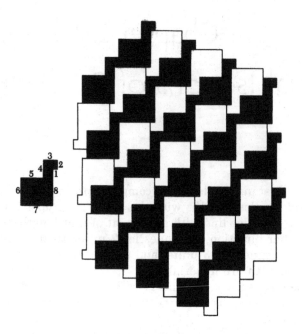

Figure 1. The only known serial-sided polyomino that tiles the plane.

It is easy to prove that N must be a multiple of 4. One way is to consider N rook moves on a bicolored chessboard. Assume without loss of generality that the rook begins on a black cell and makes its first move horizontally. To close the path, its final move must be vertical, and end on a black square. Because moves of odd and even length

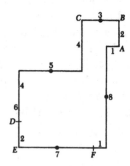

Figure 2. Applying Conway's criterion to prove that the polyomino tiles the plane.

alternate, the sequence of colors at the end of each segment forms the repeating sequence: $WWBBWWBB\ldots$. The rook returns to a black cell, after a vertical move, if and only if the number of moves $\equiv 0$ mod 4.

Experimenting on graph paper will quickly convince you that there is no serial isogon of order 4. (You might call this "a 4-gon conclusion".) We have exhibited one of order 8. After the computer program exhaustively plotted all serial isogons of orders through 24, a surprising fact emerged. No serial isogons were found except when N is a multiple of 8. This led to several proofs of the following theorem:

For any 90-degree serial isogon, N must be a multiple of 8.

Assume that a closed path begins with a unit move to the east, and that moving east or north is positive, and moving south or west is negative. A path can be uniquely described by placing a plus or minus sign in front of each number in the sequence of moves to indicate the direction of the move. For example, the order-8 polyomino has the following formula:

$$+1 + 2 - 3 - 4 - 5 - 6 + 7 + 8.$$

It is obvious that if the path closes, the sum of all horizontal moves—the odd numbers—must be zero, otherwise the path will not return to the vertical lattice line that goes through the starting point. Similarly, the sum of all vertical moves—even numbers—must be zero or the path will not return to the horizontal lattice line going through the origin point. The sum of all the numbers will, of course, also be zero.

We know that N is a multiple of 4, say $4k$. Then the north–south moves are the even-length ones, $2, 4, \ldots, 4k$. The total north–south distance is therefore $2(1 + 2 + \cdots + 2k) = 2k(2k + 1)$. Half of this, $k(2k + 1)$, must be north and half of it south. But if k is odd, this distance is odd, and cannot be the sum of even-length moves.

We can make this clearer by taking $N = 12$ as an example. Even numbers in this path's formula $(2, 4, 6, 8, 10, 12)$ add to 42. If the formula describes a closed path, the sum of the positive numbers in this sequence must equal $42/2 = 21$. But no set of even numbers can add to the odd number 21. Consequently, no formula can be constructed that will describe a closed path of order 12.

With reference to the grid, this tells us that if N is a multiple of 4 but not of 8, the last segment of the path, which is vertical, cannot return to the horizontal lattice line that goes through the path's origin

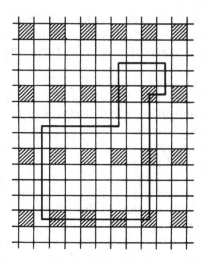

Figure 3. A bicoloring of the grid for proving that N is a multiple of 8 for all orders of closed serial path. The order-8 polyomino is shown in outline.

point because the sum of the positive segments going north cannot equal the absolute sum of the negative segments going south. The path's end will always be an even number of units above or below the zero horizontal line.

Is there a coloring pattern that proves the $N \equiv 0$ mod 8 theorem? Yes, the simple coloring shown in Figure 3 (found by the second author) will do the trick.

Start a path on any black cell, then make your first move horizontally in either direction. Regardless of your choices of how to turn at the end of each segment, the colors at the ends of segments will endlessly repeat the sequence: *WWWWWWBB, WWWWWWBB, WWWWWWBB,* .... The path will enter a black cell (which it must do if the path closes) in a direction perpendicular to the first segment if and only if the number of moves is a multiple of 8. Shown in the illustration is the path that outlines the order-8 tiling polyomino.

Although all closed paths have formulas in which the total sum of the signed terms is zero, this is not sufficient to produce a formula for a path. A serial path closes if and only if the even numbers in its formula add to zero, and likewise the odd numbers. (Nonserial closed paths may meet this proviso, and be of any order of 4 or greater that is a multiple of 2. The formula $+1 + 2 - 1 - 2$, for example, describes a closed path that outlines a domino.) If this proviso is not met, the formula gives the location of the path's final point with respect to the origin. If the sum of the signed even numbers is positive, it gives the number of units where the path ends above the origin; if negative, it gives the number of units below the origin. Similarly for the sum of the signed odd numbers. If positive, it gives the end point's distance east of the origin; if negative, the distance west.

If the signs of all even numbers are changed, or if the signs of all odd numbers are changed, it reflects the isogon along an orthogonal axis. If all signs are changed, it rotates the isogon 180 degrees.

We have shown that $N \equiv 0$ mod 8 is a necessary condition for a closed serial path. Is it also sufficient? Yes. Here is one way to arrange plus and minus signs in a formula that will always describe a serial isogon: Put plus signs in front of the first and last pairs of numbers. Put minus signs in front of the next to last pairs of numbers at each end, and continue in this way until all pairs of numbers are signed. This ensures that all even numbers add to zero, and likewise all odd numbers, therefore the formula must describe a closed path. It produces, for instance, the unique formula for $N = 8$. Applied to $N = 16$ it gives the formula $+1 + 2 - 3 - 4 + 5 + 6 - 7 - 8 - 9 - 10 + 11 + 12 - 13 - 14 + 15 + 16$, which describes the isogon at position $O_2 E_5$ in Figure 4.

Is there a procedure guaranteed to construct a serial isogon for any order $N \equiv 0$ mod 8 that outlines a polyomino? The answer is again yes. Each formula has $8n$ numbers. If we make positive all numbers in the first fourth of the formula, and also in the last fourth, and make negative all numbers in the half in between, we produce a serial isogon. Applied to $N = 16$, it gives $+1 + 2 + 3 + 4 - 5 - 6 - 7 - 8 - 9 - 10 - 11 - 12 + 13 + 14 + 15 + 16$, which describes the polyomino at $O_1 E_4$ in Figure 4. The polyomino generated by this procedure always takes the form of a snake that grows longer as N increases. Figure 5 shows the polyomino snake for order 32. A diagonal line, running from the extreme corners of the snake's head and tail, going between all the interior corners, is almost, but not quite, straight.

We now turn to the more difficult task of enumerating all possible serial isogons (not counting rotations and reflections as different) for a given isogon. As mentioned earlier, there is only one isogon of order 8, the tiling polyomino. For $N = 16$, the computer program found the 28 solutions shown in Figure 4. Note that only three ($O_1 E_1$, $O_1 E_4$, $O_1 E_7$) are polyominoes. For $N = 24$, the program produced 2,108 distinct

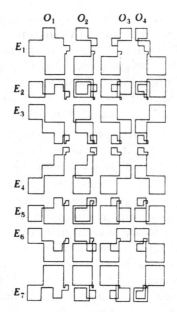

Figure 4. The 28 distinct order-16 serial isogons of $90°$.

isogons, of which 67 bound polyominoes. For $N = 32$ the program's running time became too long to be feasible.

No formula is known for enumerating all distinct serial isogons of order N, or for counting the polyominoes of a given order. However, there are procedures by which the number of isogons can be counted by hand to a value of N that goes well beyond $N = 24$.

Here is how the fourth author describes one such procedure:

Suppose $N = 16$; it's easy to see that this quantity is the constant term if you expand the algebraic product

$$(x^{-1} + x^1)(y^{-2} + y^2)(x^{-3} + x^3) \cdots (x^{-15} + x^{15})(y^{-16} + y^{16})$$

into powers of x and y. To get the number of ways for the odd sum to cancel, we want the constant term of

$$(x^{-1} + x^1)(x^{-3} + x^3) \cdots (x^{-15} + x^{15}),$$

which is the coefficient of $x^{1+3+5+\cdots+15} = x^{64}$ in

$$(1 + x^2)(1 + x^6) \cdots (1 + x^{30}),$$

which is the coefficient of x^{32} in

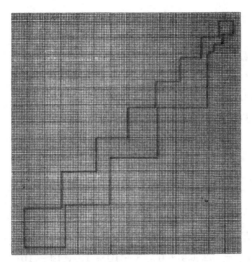

Figure 5. The snake polyomino of order 32. The diagonal line separating the snake's two sides is almost, but not quite, straight.

$$(1+x)(1+x^3)\cdots(1+x^{15}),$$

which is 8. To get the number of ways for the even sum to cancel, we want the constant term of

$$(y^{-2}+y^2)(y^{-4}+y^4)\cdots(y^{-16}+y^{16}),$$

which is the constant term of

$$(y^{-1}+y^1)(y^{-2}+y^2)\cdots(y^{-8}+y^8),$$

which is the coefficient of $y^{1+2+\cdots+8}=y^{36}$ in

$$(1+y^2)(1+y^4)\cdots(1+y^{16}),$$

which is the coefficient of y^{18} in

$$(1+y)(1+y^2)\cdots(1+y^8),$$

which is 14. So the total number of closed paths is 8×14; divide by 4 to get 28 closed paths that are distinct under reflectional symmetry. Suppose we start with $+1+2$; then the four ways to do the odd numbers are

$$O_1 = +1 + 3 - 5 - 7 - 9 - 11 + 13 + 15$$
$$O_2 = +1 - 3 + 5 - 7 - 9 + 11 - 13 + 15$$
$$O_3 = +1 - 3 - 5 + 7 + 9 - 11 - 13 + 15$$
$$O_4 = +1 - 3 - 5 + 7 - 9 + 11 + 13 - 15$$

and the seven ways to do the even numbers are

$$E_1 = +2 + 4 + 6 + 8 - 10 - 12 - 14 + 16$$
$$E_2 = +2 + 4 + 6 - 8 + 10 - 12 + 14 - 16$$
$$E_3 = +2 + 4 - 6 + 8 + 10 + 12 - 14 - 16$$
$$E_4 = +2 + 4 - 6 - 8 - 10 - 12 + 14 + 16$$
$$E_5 = +2 - 4 + 6 - 8 - 10 + 12 - 14 + 16$$
$$E_6 = +2 - 4 - 6 + 8 + 10 - 12 - 14 + 16$$
$$E_7 = +2 - 4 - 6 + 8 - 10 + 12 + 14 - 16$$

The three serial polyominoes are the snake O_1E_4 and two other solutions O_1E_1, O_1E_7. (It's curious that only O_1 can be completed. The case O_2E_1 almost works, but that path gives a degenerate polyomino whose width is zero at one point. Paths O_3E_1, O_3E_4, and O_4E_4 fail in the same way.)

In general when $N = 8n$, the number of closed paths is the product of the coefficient of x^{8n^2} in

$$(1 + x)(1 + x^3)(1 + x^5) \cdots (1 + x^{8n-3})(1 + x^{8n-1})$$

and the coefficient of y^{4n^2+2} in

$$(1 + y)(1 + y^2)(1 + y^3) \cdots (1 + y^{4n-1})(1 + y^{4n}).$$

These numbers, for small n (divided by 2 to remove symmetry), are

N	n	odds/2	evens/2	product
8	1	1	1	1
16	2	4	7	28
24	3	34	62	2108
32	4	346	657	227322
40	5	3965	7636	30276740
48	6	48396	93846	4541771016
56	7	615966	1199892	739092675672
64	8	8082457	15796439	127674038970623

It seems certain that the vast majority of these isogons will not bound polyominoes. The paper of Bhattacharya and Rosenfeld [3] is concerned with the problem of avoiding self-intersections in isogons: They treat the general problem in which the sides are of arbitrary length, not just our particular case of consective integers.

Here is how the third author has made an asymptotic estimate of the number of serial isogons:

We have seen that the number of isogons of a given order is the product of half the number of possible choices of sign in $\pm 2 \pm 4 \pm 6 \pm \cdots \pm (8n - 2) \pm 8n = 0$ with half the number of choices of sign in $\pm 1 \pm 3 \pm 5 \pm \cdots \pm (8n - 3) \pm 8n = 0$.

The first of these is the number of partitions of half of the sum

$$2 + 4 + 6 + \cdots + 8n$$

into distinct even parts, of size at most $8n$, i.e., the number of partitions of $n(4n + 1)$ into d distinct parts of size $\leq 4n$. Subtract $1, 2, ..., d$ from these parts, now no longer necessarily distinct, of size $\leq 4n - d$. We require the number of partitions of $4n^2 + n - (1/2)d(d + 1)$ into at most d parts, no longer necessarily distinct, of size $\leq 4n - d$. Here d, the number of parts, lies in the approximate range $(4 - 2\sqrt{2})n < d < 2\sqrt{2}n$.

In the same way the second number is the number of partitions of $4n^2 - (1/2)d^2$ into at most d parts, not necessarily distinct, of size $\leq 4n - d$, with d in approximately the same range as before, but with d necessarily even!

The main contribution comes from $d = 2n$ and the distribution, as we shall see, is essentially the binomial distribution, so that a good estimate of the whole is obtained by multiplying this central term by $\sqrt{4\pi n}$, except that we halve the "odd" estimate since only alternate terms (d even) are taken.

From formula (75) in [2] we learn that the number of partitions of j into at most a parts, with each part $\leq b$, is asymptotically equal to

$$\frac{1}{\sigma_{a,b}} \binom{a+b}{a} \phi\left(\frac{j - ab/2}{\sigma_{a,b}}\right),$$

where $\sigma_{a,b}^2 = ab(a+b+1)/12$ and $\phi(x) = e^{-x^2/2}/\sqrt{2\pi}$.

In both the even and the odd cases, $a = d$, $b = 4n - d$, $a + b + 1 = 4n + 1$, and the central term is given by $d = 2n = a = b$ for which $j = 4n^2 + n - n(2n+1)$ and $j = 4n^2 - 2n^2$, i.e., $2n^2$ in either case, and $j - ab/2 = 0$, so that the central term is asymptotically equal to

$$\frac{1}{\sqrt{2\pi}} \frac{\sqrt{3}}{2n\sqrt{4n+1}} \binom{4n}{2n},$$

i.e., asymptotically equal to

$$\frac{\sqrt{3}}{2n\sqrt{2\pi}\sqrt{4n+1}} \frac{2^{4n}}{\sqrt{4\pi n}}$$

by Stirling's formula. Multiply by $\sqrt{4\pi n}$ to estimate the total number of partitions in the even case, and by half of that in the odd case. The product of the halves of these numbers (i.e., not counting the E.–W. or N.–S. reflections of the isogons as different) is thus

$$\frac{1}{8}\left\{\frac{\sqrt{3}}{2n\sqrt{2\pi}\sqrt{4n+1}} 2^{4n}\right\}^2 = \frac{3 \cdot 2^{8n-6}}{\pi n^2 (4n+1)}.$$

Compare this estimate with the actual values obtained above.

n	$3 \times 2^{8n-6}/\pi n^2(4n+1)$
1	0.76
2	27.2
3	2140
4	235604
5	31248698
6	4666472281
7	756618728785
8	130321844073100

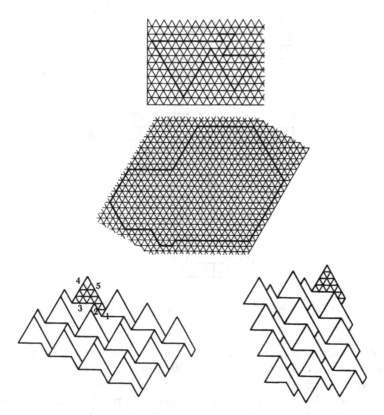

Figure 6. The unique, smallest examples of serial isometric isogons. At top is the only order-9 isogon with 60-degree angles. In the middle is the only order-12 isogon with 120-degree angles. At bottom, shown tiling the plane in two ways, is the only order-5 isogon that mixes the two angles.

The first author has since explored serial isogons on isometric grids. They are of two types: those with 60-degree angles, and those with 120-degree angles. In the 60-degree case, serial isogons exist if and only if $N \geq 9$ and $N \equiv 1 \bmod 3$. In the 120-degree case, they exist if and only if N is a multiple of 6. If each angle of a serial polygon can be either 60 or 120 degrees, such polygons exist for any N that is 5 or greater. In all three cases, the smallest order is unique and is a *polyiamond*—a polygon formed by joining unit equilateral triangles. The three are shown in Figure 6. Note that the order-5 polyiamond tiles the plane in two different ways, see [4].

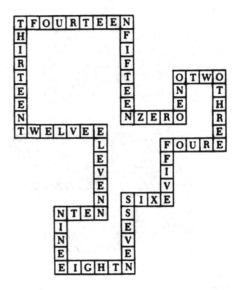

Figure 7. A 16-sided polyomino, its sides determined by the words for zero through 15.

For a time the first author believed that closed serial paths existed only when angles are 60, 90, or 120 degrees, but then he suddenly discovered such a path with 108-degree angles, the angles of a regular pentagon. More recently Hans Cornet, a retired mathematician in The Hague, has put forward a proof that at least one serial isogon can be constructed for every angle a that is a rational multiple of 360 degrees, that is, when $\alpha = 2\pi n/m$ radians, with m and n positive integers; see [4].

The first author has also investigated closed paths on square and isometric grids that have segment lengths in sequences other than the counting numbers, such as Fibonacci sequences, consecutive primes, and so on. He has produced a whimsical class of *piominoes*—polyominoes whose sides in cyclic order are the first n digits of π with zeros omitted. Because the digits of π are pseudo-random, the task of enumerating π-isogons of 90 degrees is related to determining the probability of self-intersecting random walks on a square lattice. He has even experimented with closed paths based on the letters of number words. An example is shown in Figure 7. It is unquestionably one of the most useless polyomino outlines ever constructed, yet does it not have a curious charm?

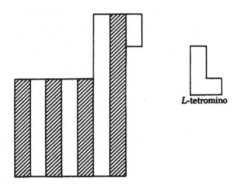

Figure 8. A coloring pattern on the tilling poliomino for proving it cannot be dissected into 13 L-tetrominoes.

Solution to Puzzle: To prove that 13 L-tetrominoes will not tile the polyomino, divide the polyomino into alternatively colored vertical stripes as shown in Figure 8. No matter how an L-tetromino is placed within this pattern, it will cover an odd number of cells of each color. Thirteen odd numbers add to an odd number, but the polyomino has an even number of cells of each color. The tiling is therefore impossible.

Postscript

The appearance of a prepublication version of this paper [4] resulted in a good deal of correspondence and misunderstanding. John Leech observed that the argument there, showing that N is a multiple of 8, is faulty and exacerbated by the statement that "even numbers hardly entered into the discussion," whereas we know that it is the even lengths that clinch the matter. Leech also gives an argument involving those gridlines that "quarter" a tetromino, which is equivalent to that of Figure 8. The difficulty of printing a complicated formula in a popular article resulted in an oversimplification that was no longer an asymptotic formula in the sense described. Finally, it was implied that calculation of the exact numbers of serial isogons was more difficult than is actually the case, which prompted several people to reach for their computers. Ilan Vardi [6] has explained how to compute the numbers rapidly via the Chinese Remainder Theorem, and he lists the values for $N = 400$ and $N = 1000$. Calculations up to $N = 200$ were carried out by Sivy Farhi, and by Pierre Barnouin, who gave the following values:

16	28
24	2108
32	227322
40	30276740
48	4541771016
56	739092675672
64	127674038970623
72	23085759901610016
80	4327973308197103600
88	835531767841066680300
96	165266721954751746697155
104	33364181616540879268092840
112	6854017416098227836106023048
120	1429368258586343246184813682344
128	302023498629081603279538134332922
136	64557914743374337032608546756101824
144	13941247125893997584457711273087122310
152	3038225349257507092516361163813831321438
160	667575475791956832191676953455074834982100
168	147773788473936923724715382248726990582150405
176	32931659242107964657022264548538525142956914056
184	7383987729780296585063944629621065123001927478725
192	1664961555710273709724126262313969341483976633058164
200	377359709872056562198423857053288570232577607987443492

References

[1] Doris Schattschneider, "Will It Tile? Try the Conway Criterion!" *Mathematics Magazine* 53 (1980), 224–232.

[2] Lajos Takács, "Some Asymptotic Formulas for Lattice Paths," *Journal of Statistical Planning and Inference* 14 (1986), 123–142.

[3] Prabir Bhattacharya and Azriel Rosenfeld, "Contour Codes of Isothetic Polygons," *Computer Vision, Graphics, and Image Processing* 50 (1990), 353–363.

[4] Lee Sallows, "New Pathways in Serial Isogons," *The Mathematical Intelligencer* 14. (1992), 56–67.

[5] A. K. Dewdney, "Mathematical Recreations: An Odd Journey Along Even Roads Leads to Home in Colygon City," *Scientific American* 263 #1 (July 1990), 118–121.

[6] Ilan Vardi, *Computational Recreations in Mathematics*, Addison-Wesley, Reading, MA, 1991, Section 5.3.

Chapter 30
Around the Solar System

In this article we examine a fantastic mathematical magic trick. I will present it as a puzzle—why does it always work?—but of course you can demonstrate it to friends as an amazing feat of ESP.

To the audience, this is how the trick appears. While your back is turned, someone is asked to put a dime on any of the nine squares shown on the facing page. Without turning around, you give instructions for moving the dime about at random over the matrix, as if it were a spaceship touring the solar system. As these random moves are made, you keep blocking off certain cells by directing that pennies be placed on them. Finally, eight cells are occupied by pennies. With your back still turned, you can name the planet on which the "spaceship" came to rest.

Pause at this point and see that you have on hand a dime and eight pennies. Instead of pennies you can use buttons, checkers, or anything else that will serve as counters. I will now assume the role of magician while you assume the role of spectator.

Select any one of the nine cells and put the dime on it. This is a completely free choice on your part, and obviously I have no way of knowing what choice you made. When you move the dime according to my instructions, you must move it one cell at a time in any horizontal or vertical direction. No diagonal moves are allowed. At each move you spell a letter in the name on the cell where you first put the dime. For example, if you start on Mars you spell M-A-R-S, moving the dime one square east, west, north, or south at random, one move for each letter.

When you finish spelling the name on the starting square, put a penny on Venus. I am, of course, betting that no matter where you began, or how you moved the dime, it will not have come to rest on Venus.

This article is reprinted from my *Riddles of the Sphinx* (1988).

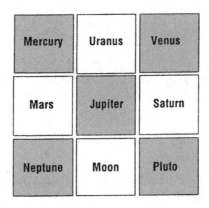

Figure 1.

From now on, at each step of your "tour" of the solar system, you move the "spaceship" just seven times, regardless of the name on the cell. These moves are made randomly, as before, but are confined to unoccupied cells. The number of these vacant squares will become fewer and fewer as more and more pennies go on the matrix.

After making seven moves, put a penny on Mars.

Move seven times. Put a penny on Mercury. As New York City's former mayor Ed Koch likes to say, "How'm I doin'?" Are all the pennies landing on vacant cells?

Move seven times. Put a penny on Uranus.

Move seven times. Put a penny on Neptune.

Move seven times. Put a penny on Saturn.

Move seven times. Put a penny on Jupiter.

Move seven times. Put a penny on the Moon.

If you followed instructions correctly, the dime should now be on Pluto!

When you show the trick to someone, turn your back while you give the above instructions. If you like, you can add to the mystery by allowing the spectator, at any time, to move by spelling S-E-V-E-N instead of counting seven. After he puts a penny on the moon, you can tell him, without turning around to look, that his dime is on Pluto.

Why does the trick always work? The answer will introduce you to the concept of "parity." It is a concept of enormous importance both in combinatorial mathematics and in modern particle physics.

Answer: Note that the names on the shaded cells have an odd number of letters and those on the unshaded cells have an even number. Mathematicians say that the two sets of squares, along with their names, are of opposite parity. One has even parity, the other odd. Each time the dime moves to an adjacent cell it changes parity.

If you start the dime on any cell and move to spell the letters of the name on that cell, the coin is sure to end on an unshaded cell. Because the dime has now acquired even parity, all shaded cells must be unoccupied, so it is safe to request that a penny be placed on shaded Venus.

From now on, at each step, the dime moves an odd number of times. It makes no difference whether it is moved seven times (or any other odd number), or moved to spell S-E-V-E-N, or any other word with an odd number of letters. If some spectator's first or last name has an odd number of letters, you can use his or her name for spelling. Every time the dime is moved, its parity alters. This allows you to direct that a penny be placed on a vacant cell of a parity opposite that of the dime. After eight steps, the only unoccupied cell will be the moon, with the dime resting on Pluto.

If you want to repeat the trick with a different final result, you'll have to work out a different set of instructions. With suitable instructions you can, of course, cause the dime to end on any of the shaded squares. Be careful, though, to eliminate the cells in such an order that the moving "spaceship" always has access to all the remaining unoccupied cells.

Chapter 31
Ten Amazing
Mathematical Tricks

Magicians have invented a fantastic variety of self-working mathematical tricks in which the outcome of seemingly random choices can be predicted in advance. Some of these tricks are described in my 1956 Dover paperback *Mathematics, Magic, and Mystery*, but hundreds of new mathematical tricks have been invented since that book was published. What follows is a small selection.

Readers are urged to follow the instructions of each trick carefully, then check the predictions at the end of this chapter. I will not spoil the fun by explaining why each magic trick works. If you can figure this out for yourself, you'll find the tests pleasant exercises in mathematical problem solving.

1. Twinkle Twinkle

> Twinkle, twinkle, little star,
> How I wonder what you are
> Up above the world so high,
> Like a diamond in the sky.
> Twinkle, twinkle, little star,
> How I wonder what you are.

Select any word in the first two lines of the above familiar poem. Count the number of its letters. Call this number n. Now count ahead n words, starting with 1 for the word following the word you selected. Count the number of letters in this second word. Call the number k. Count ahead k words to arrive at a third word.

This article first appeared in *Math Horizons* (September 1998).

Continue in this manner until you can't go any farther.
On what word does your count end?

2. 246913578

Enter the above strange number in your calculator. You may freely
choose to do any of the following:

Multiply the number by 2, 4, 5, 7, 8, 10, 11, 13, 16, 20, 22, 25, 26,
31, 35, 40, 55, 65, 125, 175, or 875.

Or, if you prefer, divide the number by 2, 4, 5, or 8.

After you have done one of the above, rearrange the digits of the
result in serial order from the smallest digit to the largest. Ignore any
zeros among the digits.

The result will be a number of nine digits.

What is this number?

3. Try This on a Dollar Bill

Write down the serial number of any dollar bill. Scramble the digits
any way you like—that is, mix up their order. Jot down this second
number.

Using your calculator, subtract the smaller number from the larger.
From the difference, subtract 7.

Copy the digits now on display, then add them all together. If the
sum is more than one digit, add the digits once more.

Keep adding digits in the sum until just one digit is obtained.
What is it?

4. Count the Matches

From an unused folder of 20 paper matches, tear out any number of
matches less than ten and put them in your pocket.

Count the number of matches remaining in the folder. Add the
two digits of the count, then remove that number of matches from the
folder. Put them in your pocket.

Tear out three more matches.

How many matches are left in the folder?

5. A Test with Two Dice

Roll a pair of dice on the table. Call them A and B.
Write down the following four different products:

1. The product of the top numbers on the dice.

2. The product of their bottom numbers.

3. The product of the top of A and the bottom of B.

4. The product of the top of B and the bottom of A.

Add the four products. What's the sum?

6. Fold and Trim

Fold a sheet of paper in half four times, and then unfold it. The creases will form a 4 × 4 matrix of cells as shown in Figure 1. Number the cells from 1 through 16 as illustrated. Fold each crease forward and back a few times so the paper will fold easily either way along each crease.

Now fold the sheet into a packet the size of one cell. You can make the folds as tricky as you please, folding any way you like. You may even tuck folds between folds. In other words, make the folding as random as possible until you have a packet the size of a single cell.

With scissors trim away all four edges of the packet so that it consists of sixteen separate pieces. Spread these pieces on the table.

1	2	3	4
5	6	7	8
9	10	11	12
13	14	15	16

Figure 1.

Some will have their number side up, others will have their number side down.

Add all the numbers on the face-up pieces.

What is the sum?

7. At the Apex

Copy the triangle of circles in Figure 2.

Put any four digits you like in the four vacant circles of the bottom row. They needn't be all different, and you may include one or more zeros if you like.

The remaining circles are filled with digits as follows: Add two adjacent pairs of numbers, divide the sum by 5, and put the remainder in the circle just above the adjacent pair of numbers.

For example, suppose two adjacent numbers in the bottom row are 6 and 8. They add to 14. Dividing 14 by 5 gives a remainder of 4, so you put 4 directly above the 6 and 8. If there is no remainder (such as 6 + 4 = 10) then put a zero above the 6 and 4.

Continue in this way, going up the triangle, until all the circles have digits.

What digit is at the apex of the triangle?

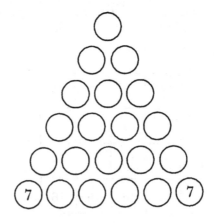

Figure 2.

8. The Red and the Black

Shuffle a deck of cards, then deal 30 cards to the table to form a pile.

Count the number of black cards in the pile. From this number subtract the number of red cards in the rest of the deck.

What's the difference?

9. Four File Cards

Write 39 on one side of a file card, and 51 on the other side. On a second file card write 26 and 34 on the two sides. A third card gets 65 and 85. A fourth card gets 52 and 68.

Place the four cards on the table so the numbers showing on top are 26, 39, 52, and 65.

Slide any card out of the row, then turn over the three remaining cards. Slide out another card, and turn over the remaining two cards.

Slide out a third card. Turn over the card that remains.

You now have a choice of leaving the cards as they are, or turning all of them over.

With your calculator, multiply all the numbers showing. What is the product?

10. A 3 × 4 Test

Copy the 3 × 4 matrix in Figure 3.

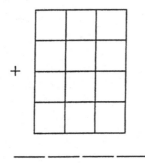

Figure 3.

Put digits 1 through 9 in the cells in any way you like. Three empty cells will remain. In those cells put either three ones, or three fours, or three sevens.

Treat each row of the matrix as a 3-digit number. Add these four numbers by writing the sum over the four lines below the matrix.

Add the sum's four digits. If the result is more than one digit, add those two numbers. Keep adding until only one digit remains.

What is this digit?

Predictions

1. The count ends on the word "you." The trick is based on what magicians call the Kruskal Count, a principle discovered by Princeton mathematician Martin Kruskal.

2. The number is 123456789.

3. The final digit is 2.

4. Six matches remain in the folder.

5. The sum of the four products is 49.

6. The face-up numbers add to 68.

7. The triangle's top number is 4.

8. The difference between the black and red cards is 4.

9. The product of the four numbers is 5860920.

10. The final digit is 3.

Chapter 32
Modeling Mathematics with Playing Cards

Because playing cards have values 1 through 13 (jacks 11, queens 12, kings 13), come in two colors, four suits, and have fronts and backs, they provide wonderfully convenient models for hundreds of unusual mathematical problems involving number theory and combinatorics. What follows is a choice selection of little-known examples.

One of the most surprising of card theorems is called the Gilbreath principle, after magician Norman Gilbreath who first discovered it. Arrange a deck so the colors alternate. Cut it so the bottom cards of each half are different colors, then riffle shuffle the halves together. Take cards from the top in pairs. Amazingly, every pair will consist of a red and black card!

Here is a simple proof by induction. Assume that the first card to fall on the table during the shuffle is black. If the next card to fall is the card directly above it in the same half, that card will be red. This places on the table a red/black pair. If the next card after the first one comes from the other half, it too will be red to put a red/black pair on the table. In either case, after two cards have dropped, the bottom cards of each half will be of different colors, so the situation is exactly the same as before and the same argument applies for the rest of the cards. No matter how careful or careless the shuffle, it will pile red/black pairs on the table.

Gilbreath's principle generalizes. Arrange the deck so the suits are in an order, say spades, hearts, clubs, diamonds, that repeats throughout the pack. Deal as many cards as you like to form a pile. This of course reverses the order of the suits. When the pile is about the same size as the remaining portion of the deck, riffle shuffle the

This article first appeared in *The College Mathematics Journal* (Vol. 31, May 2000).

two portions together. If you now take cards in quadruplets from the top of the shuffled pack you will find that each set of four contains all four suits.

The ultimate generalization is to shuffle together two decks, one with its cards in the reverse order of the other deck. After the shuffle, divide the 104-card pack exactly in half. Each half will be a complete deck of 52 different cards!

What mathematician David Gale has called the "non-messing-up theorem" is another whimsical result. From a shuffled deck deal the cards face up to form a rectangle of any proportion. In each row re-arrange the cards so their values do not decrease from left to right. In other words, each card has a value higher than the one on its left, or two cards of the same value are side by side.

After ordering the rows, do the same thing with the columns. This of course drastically alters the order of cards in the rows. After re-arranging the columns, you may be amazed to find that the rows are still ordered!

The theorem is at least a hundred years old. You will find it proved as the answer to a problem in *American Mathematical Monthly,* Vol. 70, February 1963, pp. 212–13, and in a monograph by Gale and Richard Karp, published in 1971 by the operations research center of the engineering school of the University of California, Berkeley. Donald Knuth discusses the theorem in the third volume of *The Art of Computer Programming* in connection with a method of sorting called "shellsort." In my *The Last Recreations*, Chapter 11, I describe a clever card trick based on the theorem.

Is it possible to arrange a deck so that if you spell the name of each card by moving a card from top to bottom for each letter, then turning over the card at the end of the spell and discarding it, it will always be the card you spelled? For example, can you so arrange the cards that you can first spell all the spades, taking them in order from ace through king, then do the same thing with the hearts, clubs, and diamonds?

You might imagine it would take a long time to find out how to arrange the deck, assuming it is possible to do so, in a way that permits the spelling of all 52 cards. Actually, finding the order is absurdly easy. First arrange the deck from top down in the order that is the reverse of your spelling sequence. Take the King of Diamonds from the top of the deck, then take the queen, place it on top of the king, and spell "Queen of Diamonds" by moving a card at each letter from bottom to top. In brief, you are reversing the spelling procedure. Continue in this way until the new deck is formed. You are now all set to spell every card in the predetermined order. Of course you can do the same thing with smaller packets, such as the thirteen spades, or with cards bearing pictures, say of animals whose names you spell.

Remember the old brain teaser about two glasses, one filled with water, the other with wine? You take a drop of water, put it into the wine, stir, then take a drop of the mixture, move it back to the water, and stir. Is there now *more* or *less* wine in the water than water in the wine?

The answer is that the two quantities are exactly equal. The simplest proof is to realize that, after the transfers, the amount of liquid in each glass remain the same. So the quantity missing from the water is replaced by wine, and amount of wine missing from the other glass is replaced by the same amount of water.

This is easily modeled with cards. Divide the deck into two halves, one of all red cards, the other of all black. Randomly remove n red cards, insert them anywhere in the black half, and shuffle. Now randomly remove n cards from the half you just shuffled, put them back among the reds, and shuffle. Inspection will show that the number of black cards in the red half exactly equals the number of red cards in the black half. It doesn't matter in the least if the red and black portions are not equal at the start.

Closely related to this demonstration is the following trick. Cut a deck exactly in half, turn over either half and shuffle the two parts together. Cut the mixed-up deck in half again, and turn over either half. You'll find that the number of face-down cards in either half exactly equals the number of face-down cards in the other half. The same is true, of course, for the face-up cards. Do you see why this is the case? The trick is baffling to spectators if they don't know that the deck is initially divided exactly in half, and if you secretly turn over one half as you spread its cards on the table.

Playing cards provide a wealth of counter-intuitive probability questions. A classic instance involves three cards that are face down on the table. You are told that one card only is an ace. Put a finger on a card. Clearly the chances you have selected the ace is 1/3. A friend now secretly peeks at all three cards and turns face up a card that is not the ace. Two cards remain face down, one of which you know is the ace. What now is the probability your finger is on the ace? Many persons, including mathematicians who should know better, think the probability has risen from 1/3 to 1/2. A little reflection should convince you that it remains 1/3 because your friend can always turn a non-ace.

Now switch your finger from the card it is on to the other face-down card. It may be hard at first to believe, but the probability you have now chosen the ace jumps from 1/3 to 2/3! This is obvious from the fact that the card you first selected has the probability of 1/3 being the ace. Because the ace must be one of the two face-down cards, the two probabilities must add to 1, or certainty.

A similiar-seeming paradox also involves three face-down cards dealt from a shuffled deck. A friend looks at their faces and turns over two that are the same color. What's the probability that the re-maining face-down card is the same color as the two face-up cards? You might think it is 1/2. Actually it 1/4. Here's the proof. The proba-bility that three randomly selected cards are the same color is two out eight equal possibilities, or 1/4. Subtract 1/4 from 1 (the card must be red or black) and you get 3/4 for the probability that the face-down card differs in color from the two face-up cards. This is the basis for an ancient sucker bet. If you are the operator, you can offer even odds that the card is of opposite color from the two face-up cards, and win the bet three out of four times.

Here's a neat problem involving a parity check. Take three red cards from the deck. Push one of them back into the pack and take out three black cards. Push one of them back into the deck and remove three reds. Continue in this manner. At each step you randomly select a card of either color, return it to the deck and remove three cards of opposite color. Continue as long as you like. When you decide to stop you will be holding a mixture of reds and blacks. Is it possible that the number of black cards you hold will equal the number of red? Unless you think of a parity check it might take a while to prove that the answer is no. After each step you will always have in your hand an odd number of cards, therefore the two colors can never be equal.

Magicians have discovered the following curiosity. Place cards with values ace through nine face down in a row in counting order, ace at the left. Remove a card from either end of the row. Take another card from either end. Finally, take a third card from either end. Add the values of the three cards, then divide by six to obtain a random number n. Count the cards in the row from left to right, and turn over the nth card. It will always be the four!

I leave it to readers to figure out why this works and perhaps to generalize it to longer rows of numbers. For example, use twelve cards with values 2, 3, 4, 5, 6, 7, 8, 9, 10, J, Q, K to make the row. Take a card three times from either end, divide the sum of the values by 9, and call the result n. The nth card from the left will always be the five.

A classic card task, going back more than two centuries, is to arrange all the aces, kings, queens, and jacks—sixteen cards in all—in a square array so that no two cards of the same value, as well as no two cards of the same suit, are in the same row, column, or diagonal.

Counting the number of different solutions is not trivial. W.W. Rouse Ball, in his classic *Mathematical Recreations and Essays*, said there are 72 fundamental solutions, not counting rotations and reflections. This was a mistake that persisted through the book's eleventh edition, but was dropped from later editions revised by H.S.M. Coxeter. Dame Kathleen Ollerenshaw, a noted British mathematician who was once Lord Mayor of Manchester, found there are twice as many fundamental solutions, 144, making the number of solutions including rotations and reflections $8 \times 144 = 1152$. She recently described a simple procedure for generating all 1152 patterns in an article written for the blind. (Dame Ollerenshaw, now 87, is slowly losing her vision, and energetically learning how to read Braille.)

This is her procedure: Number the sixteen positions in the array from 1 through 16, left to right, top down. Place an arbitrary card, say the Ace of Spades, in position 1, the top left corner. A second ace, say the Ace of Hearts, goes in the second row. It can't go in the same column or diagonal as the Ace of Spades, so it must go in either space 7 or 8. Place it arbitrarily in space 7.

Two aces remain to go in rows 3 and 4. Put the Ace of Diamonds in the third row. It can go only in space 12. The Ace of Clubs is now forced into space 14 of the bottom row. Had the second ace gone in space 8, the last two aces would have been forced into spaces 10 and 15.

Consider the other three spades. They can't go in the top row or leftmost column, or in a main diagonal. This forces them into spaces 8, 10, and 15. Arbitrarily place the King of Spades in 8, the Queen of Spades in 10, the Jack of Spades in 15. The pattern now looks like this:

A♠			
		A♥	K♠
	Q♠		A♦
	A♣	J♠	

The remaining nine cards are forced into spaces that complete the following pattern:

A♠	K♥	Q♦	J♣
Q♣	J♦	A♥	K♠
J♥	Q♠	K♣	A♦
K♦	A♣	J♠	Q♥

Multiply the number of choices at each step, $16 \times 3 \times 2 \times 2 \times 3 \times 2$, and you get the total of 1152 patterns.

For more examples of mathematical theorems, problems, and tricks with playing cards, see my Dover paperback *Mathematics, Magic, and Mystery*, and Karl Fulves's *Self-Working Card Tricks*, also a Dover paperback, and the following chapters in my collections of *Scientific American* columns: *The Scientific American Book of Mathematical Puzzles and Diversions*, Chapter 10; *Mathematical Carnival*, Chapters 10 and 15; *Mathematical Magic Show*, Chapter 7; *Wheels, Life, and Other Mathematical Amusements*, Chapter 19; *Penrose Tiles to Trapdoor Ciphers*, Chapter 19, and *The Last Recreations*, Chapter 2.

Now for two puzzles that can be modeled with cards. Solutions appear at the end of the chapter.

1. Arrange nine cards as shown in Figure 1. Assume the aces have a value of 1. Each row, each column, and one diagonal has a sum of 6. The task is the alter the positions of three cards so that the matrix is fully magic for all rows, columns, and diagonals.

A	2	3
3	A	2
2	3	A

Figure 1.

2. Nine cards arranged as shown in Figure 2 has the property of minimizing the sum of all absolute differences between each pair of cells that are adjacent vertically and horizontally. Assume that the matrix is toroidal—that is, it wraps around in both directions. The sum of the differences is 48. This was proved minimal by Friend Kirstead, Jr., in the *Journal of Recreational Mathematics*, Vol. 18 (1985–86), p. 301.

A	2	3
4	5	6
7	8	9

Figure 2.

The challenge is to take nine cards of distinct values (court cards may be used) and form a toroidal square that will *maximize* the sum of all absolute differences.

Solutions to the Puzzles:

1. Move the bottom row of three cards to the top, or move the left-most column to the right, Either change produces the desired magic square:

2	3	A
A	2	3
3	A	2

2. The figure below shows how nine cards of distinct values can be placed in a toroidal square so as to maximize the sum of absolute differences of adjacent values in rows and columns. Any value 4 through 10 can go in cells x, y, and z to make a total of 120. This was proved maximal by Brian Maxwell, of Middlesex, England, in the *Journal of Recreational Mathematics*, Vol. 18 (1985–86), p. 300.

A	K	Z
Q	Y	2
X	3	J

Chapter 33

The Asymmetric Propeller Theorem

The late Leon Bankoff (he died in 1997) was a Beverly Hills, California, dentist who was also a world expert on plane geometry. (For G. L. Alexanderson's interview with Bankoff, see [1].) We became good friends. In 1979 he told me about a series of fascinating discoveries he had made about what he called the asymmetric propeller theorem. He intended to discuss them in an article, but never got around to it. This is a summary of what he told me.

The original propeller theorem goes back at least to the early 1930s and is of unknown origin. It concerns three congruent equilateral triangles with corners meeting at a point as shown shaded in Figure 1. The triangles, which resemble the blades of a propeller, need not form a symmetrical pattern, but may be in any position. They may touch one another or even overlap. Chords $\overline{BC}$, $\overline{DE}$, and $\overline{FA}$ are drawn to form a hexagon inscribed in a circle. The midpoints of the three chords mark the vertices of an equilateral triangle.

A proof of the theorem, using complex numbers, appeared in [2] as the answer to Problem B-1 in the annual William Lowell Putnam Competition. H. S. M. Coxeter sent the proof to Bankoff on a Christmas card, asking him if he could provide a Euclidian proof of the theorem.

Bankoff had no difficulty finding such a proof. In a paper titled "The Asymmetric Propeller" [3], Bankoff, Paul Erdős, and Murray Klamkin made the first generalization of the theorem. They showed that the three equilateral triangles need not be congruent. They can be of any size, as shown in Figure 2, and the theorem still holds. Two proofs are given, one a simple Euclidian proof, the other with complex

This article first appeared in *The College Mathematics Journal* (Vol. 30, January 1999, pp. 18–22).

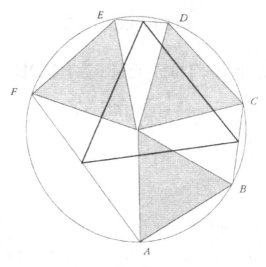

Figure 1.

numbers. As before, and in all subsequent extensions, the triangles may touch one another or even overlap.

Later, Bankoff made three further generalizations. As far as I know they have not been published.

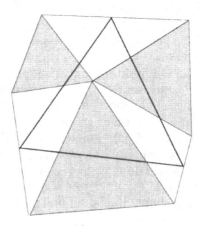

Figure 2.

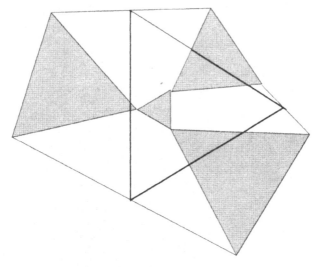

Figure 3.

Second generalization: The propeller triangles need not meet at a point. They may meet at the corners of any equilateral triangle, as shown in Figure 3.

Third generalization: The propeller triangles need not be equilateral! They need only be similar triangles of any sizes that meet at a point. The midpoints of the three added lines will then form a triangle similar to each of the propellers, as shown in Figure 4.

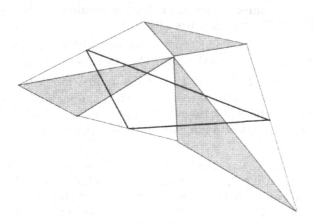

Figure 4.

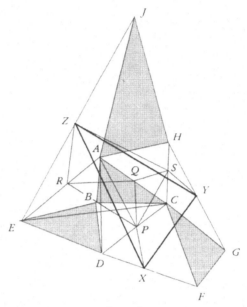

Figure 5.

Fourth generalization: The similar triangles need not meet at a point! If the propellers meet at the corners of a fourth triangle of any size, provided it is similar to each propeller, the midpoints of the added lines will form a triangle similar to each propeller. Vertices of the interior triangle must touch corresponding corners of the propellers.

Here is how Bankoff proved his final generalization on a sheet that he typed in 1973. It makes use of Figure 5.

The propellers shown are right triangles, although they can be any type of triangle. Perhaps the proof that follows can be simplified.

If ABC, AHJ, DBE, and FGC are similar triangles, all labeled in the same sense and situated so that corresponding angles meet at the vertices of triangle ABC, then X, Y, Z, the midpoints of $\overline{DF}$, $\overline{GH}$ and $\overline{JE}$, are vertices of a triangle similar to the other four.

Proof. Let $\angle BCA = \angle GCF = \alpha$; $\angle DBE = \angle ABC = \beta$; $\angle JAH = \angle CAB = \gamma$; and let P, Q, R, S denote the midpoints of the segments $\overline{DC}$, $\overline{AC}$, $\overline{AE}$ and $\overline{CH}$ respectively. We proceed stepwise to show that triangles PQR, PSZ and finally XYZ are similar to triangle ABC.

If triangle ABD is pivoted about B so that $\overline{AB}$ falls along $\overline{BC}$ and $\overline{DB}$ along $\overline{EB}$, it is seen by the relation $AB/BC = DB/BE$ and by the equality of $\angle ABD$ and $\angle CBE$ that triangles ABD and CBE are similar and that $EC/AD = BC/AB$, with $\angle EC, AD = \angle BC, BA = \beta$.

Since RQ is parallel to and equal to half EC while QP is parallel to and equal to half AD, we extend the previous relation to read $RQ/QP = EC/AD = BC/BA$, with $\angle RQ, QP = \angle EC, AD = \angle BC, BA = \beta$. It follows that triangles PQR and ABC are similar.

In like manner, because of the relationship of $\overline{AJ}$ and $\overline{AH}$ to $\overline{RZ}$ and $\overline{QS}$ as well as to $\overline{RP}$ and $\overline{QP}$ in both relative length and in direction, we find triangles ZRP and SQP similar. Then $ZP/SP = ZR/QS = AJ/AH = AC/AB$, with the angles between the segments in the numerator and in the denominator all equal to γ. As a result, triangles PSZ and ABC are similar.

Continuing as before, we find triangles ZPX and ZSY similar since $PX/SY = CF/CB = CA/CG$ and $\angle PX, SY = \angle CF, CG = \angle CA, CB = \alpha$.

Noting that in the similar triangles ZPX and ZSY we have $ZX/ZY = ZP/ZS = CA/CB$ and $\angle ZX, ZY = \angle CA, CB = \alpha$, we conclude that triangles XYZ and ABC are similar.

And now a question for interested readers to explore. Do the propellers have to be triangles? It occurred to me that if squares are substituted for triangles, as in Figure 6, that equilateral triangle still shows up.

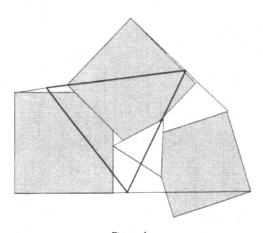

Figure 6.

I have written this piece as a tribute to one of the most remarkable mathematicians I have been privileged to know.

References

[1] G. L. Alexanderson, A conversation with Leon Bankoff, *College Mathematics Journal* 23:2 (1992) 98–117.

[2] *American Mathematical Monthly* 75:7 (1968) 732–739.

[3] Leon Bankoff, Paul Erdös, and Murray Klamkin, The asymmetric propeller, *Mathematics Magazine*, 46:5 (1973) 270–272.

Postscript

When I discussed the propeller theorem with Leon Bankoff, he showed me the following somewhat similar theorem. If two squares of arbitrary size touch at a corner, as shown in Figure 7, and their centers are joined to midpoints on lines connecting pairs of corners, the result is another square. He said he was working on a proof. I cannot now recall if the theorem was original with Leon or whether he had encountered it in a periodical.

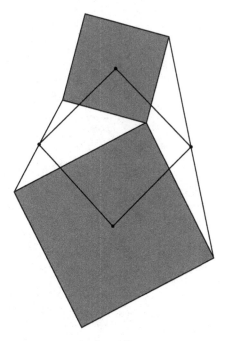

Figure 7.

Chapter 34
Chess Queens and Maximum Unattacked Cells

There is now an enormous literature on the old classic task of placing eight queens on a chessboard so that no queen can attack another. There are twelve solutions, not counting trivial rotations and reflections. The task naturally generalizes to enumerating the number of solutions for n non-attacking queens on an $n \times n$ board. (See Chapter 16 of my *Unexpected Hanging and Other Mathematical Diversions*, 1969.)

Less well known is an intriguing similar task. Let me introduce it with a difficult little puzzle that I thought of and first explained in my May 1972 *Scientific American* column, and gave again in a February 1978 column. Can you place five white queens and three black queens on a 5 × 5 board so that no queen of one color can attack a queen of another color? The only solution is shown in Figure 1. To prove it unique, the best way is to explore how three queens can be placed to leave five cells unattacked.

This suggests the following generalization: How can n queens be placed on an order-n board so as to maximize the number of unattacked vacant cells? And how many different ways can this be done for any n?

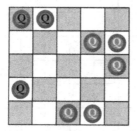

Figure 1.

This article first appeared in *Math Horizons* (November 1999).

Let's generalize further. Given a board of order n, what is the largest number of unattacked vacant cells that k queens will allow, and in each case, how many distinct patterns solve the problem? The same question can be more colorfully framed along the lines of the order-5 problem. Given an order-n board and k queens of one color, how can they be placed so as to maximize the number of queens of a different color that can be put on the board so no queen of one color can attack a queen of the other color?

The earliest publication known to me of a problem of this sort is in W. W. Rouse Ball's classic *Mathematical Recreations and Essays* (third edition, 1896). In his chapter on chess recreations Ball attributes the problem to his friend Captain W. H. Turton. The task is to place eight queens on the chessboard so as to leave unattacked the largest number of vacant cells. One solution is given. Ball comments: "Is it possible to place the eight queens so as to leave more than eleven cells out of check? I have never succeeded in doing so, nor in showing that is impossible to do it."

Henry Ernest Dudeney's Problem 316, in his *Amusements in Mathematics* (1917), involves eight queens placed on the chessboard as shown in Figure 2. The puzzle, which Dudeney says is based on Captain Turton's problem, is to alter the positions of three queens so that eleven vacant cells are unattacked. The unique solution is shown in Figure 14.

"I will hazard the statement," Dudeney writes, "that eight queens cannot be placed on the chessboard so as to leave more than eleven squares unattacked. It is true that we have no rigid proof of this yet, but I have entirely convinced myself of the truth of the statement."

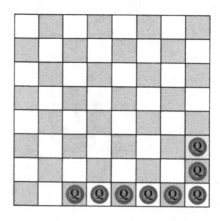

Figure 2. Dudeney's eight queens puzzle.

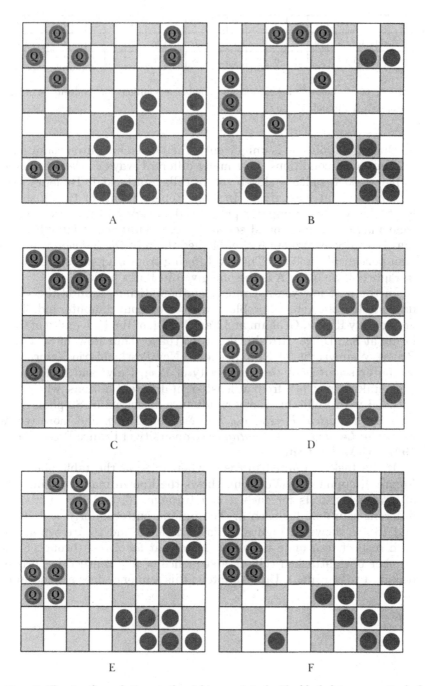

Figure 3. The six other solutions to the eight queen's task. The black dots are unattacked cells.

n	1	2	3	4	5	6	7	8	9	10	11	12	13	14	15	16
unattacked cell	0	0	0	1	3	5	7	11	18	22	30	36	47	56	72	82
number of solutions	0	0	0	25	1	3	38	7	1	1	2	7	1	4	3	1

Figure 4. Maximum unattacked cells and number of solutions for n queens on order-n boards, $n = 1$ through 16.

Dudeney's solution is unique only when given his initial conditions. Without such conditions, how many different ways can eight queens be placed so as to leave eleven vacant cells unattacked? Dudeney says he knows of "at least five." In 1995 Robert Trent, of Hardinsburg, Kentucky, wrote a computer program that verified that eleven is indeed maximum, and found seven distinct solutions. I later learned that Bernd Schwarzkopf had published them in *Die Schwalbe*, a German periodical, in 1982. The six differing from Dudeney's are shown in Figure 3. Solution A is the one given in Ball's book.

In the past few years a raft of mathematicians around the world have studied the general problem, with or without computer aid. They are notably Ronald Graham and Fan Chung, of the University of California at San Diego; Hiroshi Okuno, of Tokyo; Thur Row, of St. Louis; Robert Wainwright, of New Rochelle, New York; Mario Velucchi, a computer scientist at the University of Pisa, Italy; and mathematicians unknown to me, from at least eight different nations, who corresponded with Velucchi. Early results are discussed in Stephen Ainsley's *Mathematical Puzzles* (G. Bell & Sons, 1977), and more recent results in *Les Jeux Mathématiques* (University of France Press, Paris, 1997), by Michel Criton.

If the task is limited to cases where $k = n$, the table shown in Figure 4, supplied by Velucchi, shows the known maximum number of unattacked cells for each $n = 1$ through 16, and the number of different solutions for each n. The unique patterns for $n = 9, 10, 13$, and 16 are shown in Figure 5. Patterns for 9 and 16 are symmetric with respect to a main diagonal. Note that for $n = 4$ through 8 the number of unattacked cells is a sequence of five consecutive primes (taking 1 to be prime). Unfortunately, this sequence does not continue.

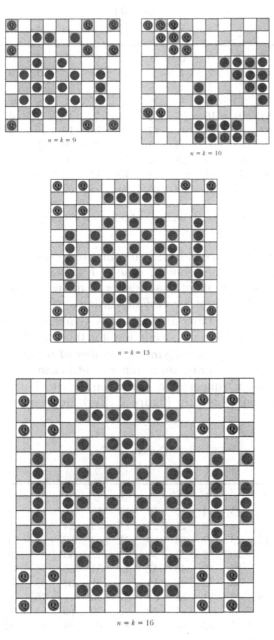

Figure 5. Unique patterns for $n = k = 9$, 10, 13, and 16.

Queens (k)

	1	2	3	4	5	6	7	8	9	10	11	12
3	2	1	0	0	0	0	0	0	0	0	0	0
4	6	3	2	1	1	1	0	0	0	0	0	0
5	12	7	5	4	3	2	2	1	1	1	1	1
6	20	13	9	8	6	5	4	4	3	2	2	2
7	30	21	16	16	11	10	7	6	6	6	5	4
8	42	31	25	24	17	15	13	11	10	9	8	7
9	56	43	36	36	27	24	20	19	18	14	13	12
10	71	57	49	48	37	33	30	26	24	22	19	18
11	90	73	64	64	51	46	42	39	36	32	30	29
12	110	91	81	80	65	59	56	50	46	42	38	36

Board (n) — rows 3 through 12

Figure 6. Maximum number of unattacked cells, n = 3 through 12, k = 1 through 12.

The chart in Figure 6, based on computer results from Okuno and Trent, covers the more general case of k queens on order-n boards. Very little is known about the number of distinct solutions for given values of k and n except for very low orders.

Now for some amazing discoveries by Graham and Chung. I publish them here, with their permission, for the first time.

Instead of listing the maximum number of unattacked cells for k queens, we list the minimum number of attacked cells for k queens, assuming that each queen attacks the cell it is on. This yields the complementary chart shown in Figure 7. Each number on this chart ob-

Queens (k)

	1	2	3	4	5	6	7	8	9	10	11	12
3	7	8	9	9	9	9	9	9	9	9	9	9
4	10	13	14	15	15	15	16	16	16	16	16	16
5	13	18	20	21	22	23	23	24	24	24	24	24
6	16	23	27	28	30	31	32	32	33	34	34	34
7	19	28	33	33	38	39	42	43	43	43	44	45
8	22	33	39	40	47	49	51	53	54	55	56	57
9	25	38	45	45	54	57	61	62	63	67	68	69
10	28	43	51	52	63	67	70	74	76	78	81	82
11	31	48	57	57	70	75	79	82	85	89	91	92
12	34	53	63	64	79	85	88	94	98	102	106	108

Board (n) — rows 3 through 12

Figure 7. Minimum number of attacked cells, n = 3 through 12, k = 1 through 12.

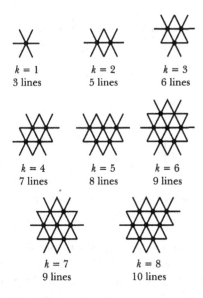

Figure 8. The number of lines generated by k spots on a hexagonal lattice correlates with the differences between consecutive values for the minimum number of attacked cells by k queens as n (a board's order) increases.

viously is obtained by subtracting each number on the previous chart from n^2, the number of squares on an $n \times n$ board.

Look down the column for $k = 1$, a single queen on an order-n board. The best spot for it is at a corner. As n increases, the values in the table have a constant difference of 3. The values in the column for two queens have a constant difference of 5. The values for three queens start with a difference of 5, followed by 6, 7, 6, 6, 6,.... The sixes continue to infinity. Indeed, Graham and Chung have proved that for every value of k, the consecutive differences, as n increases, eventually become constant!

What is more astonishing, the value of this constant, with the sole exception of $k = 4$, is exactly equal to the number of lines generated by k points that spiral around a fixed point on a hexagonal lattice? Figure 8 shows how this works for points 1 through 8. The chart shown in Figure 9 gives the value of the constant difference for each k, from 1 through 19. Note: The starred numbers are hexagonal numbers of the form $3x^2 - 3x + 1$.

k	consecutive differences	k	consecutive differences
*1	3	11	12
2	5	12	12
3	6	13	13
4	5/7	14	13
5	8	15	14
6	9	16	14
*7	9	17	15
8	10	18	15
9	11	*19	15
10	11		

Figure 9. For every k, beyond a certain value, the consecutive differences between minimum attacked cells, as n increases, become constant.

Graham and Ghung were unable to obtain a precise expression for the value of n beyond which the consecutive differences become constant. They estimate it to be within the range of $4k$ or $5k$.

In the exceptional case of $k = 4$, the consecutive constant difference, beyond $n = 5$, oscillates between 5 and 7. The minimum number of attacked cells is $6n - 8$ when n is even, and $6n - 9$ when n is odd. This arises from the fact that a best pattern for four queens, when n exceeds 4, is obtained by placing the queens in the board's four corners. When n is odd and greater than 3, the board's two main diagonals, along which the queens attack, share the board's central cell, so the formula subtracts 9 from $6n$ rather than 8 in the even case where the two main diagonals do not intersect.

Formulas for the maximum number of unattacked empty cells are known only for small k. For example, if there is only one queen, a maximum number is obtained by putting the queen in a corner. For every n the number of unattacked cells is $n^2 - 3n + 2$. If there are two queens, solutions are obtained by putting one queen in a corner, and the other in the same row or column, on the third cell from the corner occupied by the other queen. In this case the number of unattacked cells is $n^2 - 5n + 7$.

Wainwright found numerous "pretty" solutions. Figure 10, for example, is one of many symmetrical solutions he found for $n = k = 12$. Trent suggested the narrower task of finding symmetric patterns on which the number of queens equals the maximum number of unattacked cells. In such cases, of course, the queens and the cells not in check can be exchanged. Two bilaterally symmetric examples

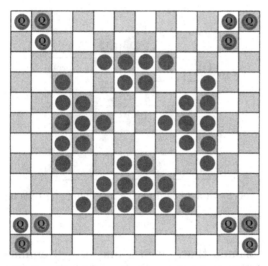

Figure 10. Wainwright's bilaterally symmetric solution for $k = n = 12$.

on order-9 boards are shown in Figure 11. Figure 12 shows how seven queens fail to attack the maximum of seven cells on an order-7 board.

Two questions remain open:

1. Find a formula that, given the number of queens, will determine the constant difference between the numbers of attacked cells as n increases beyond a certain point.

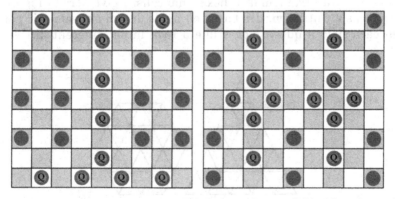

Figure 11. Two symmetric patterns from Bob Trent on the 9×9 field with k, the number of queens, equal to the maximum number of unattacked cells.

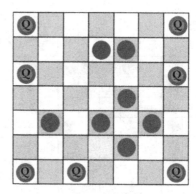

Figure 12. Seven queens on a 7-board have a maximum of 7 cells unattacked. This is one of 38 solutions.

2. Find a formula that, given k and n, will enumerate the number of different patterns that produce a maximum number of unattacked cells. This could be an enormously difficult problem in combinatorics. Moreover, there is no reason at present to assume that such a formula even exists.

I have not given any of the neat proofs by which Graham and Chung arrived at their results because they are a bit too advanced for this magazine. Perhaps the two will provide them some day in a technical paper.

Similar tasks involving rooks, bishops, and knights have been investigated by Velucchi. I know of no work done on queens that move on boards with triangular or hexagonal cells. I give here in Figure 13 two pleasant little puzzles based on fields drawn on isometric paper. On each field a queen moves like a rook in six different directions,

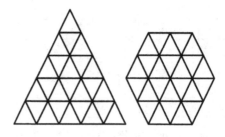

Figure 13. Two unattacked cells problems.

horizontally left or right, and up and down in two diagonal directions. In other words, a queen moves along rows of cells that are between adjacent parallel lines. Place four queens on each board maximizing the number of unattacked cells. The triangular pattern has only one solution. The hexagonal pattern has three.

Perhaps some readers will be interested in generalizing these results to larger isometric fields. Anyone wishing to correspond with Velucchi can reach him at Via Emilia, 106, 1-56121, Pisa, Italy.

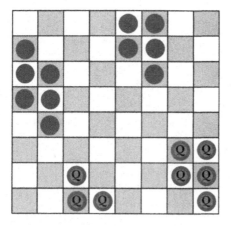

Figure 14. Dudeney's solution to his eight queen puzzle.

Part II

Chapter 35
Lion Hunting

Lion Hunting and Other Mathematical Pursuits: A Collection of Mathematics, Verse and Stories by Ralph P. Boas, Jr. Gerald L. Alexanderson and Dale H. Mugler, eds. xii + 308 pp. The Mathematical Association of America, 1995. $35.

Ralph Philip Boas, Jr., died in 1992 without seeing the publication of this splendid tribute to his distinguished career. He was not only a creative mathematician of top rank, but also an inspired teacher, long-time chairman of Northwestern University's mathematics department and one-time editor of *Mathematical Reviews* and *The American Mathematical Monthly*.

Mathematicians Alexanderson and Mugler have done an admirable job of assembling this collection of the lighter side of Boas' writings, although some of the papers may not seem "light" to readers unfamiliar with advanced mathematics. Interspersed among such serious articles are short stories, humorous verse, amusing anecdotes and colorful reminiscences.

Among Boas' popularly written papers the two best known are "Contributions to the Mathematical Theory of Big Game Hunting," and "Snowfalls and Elephants, Pop Bottles, and Pi." The first of these outlines 16 ways to capture a lion. For example, you enter a spherical cage in the desert, lock it, then perform a space inversion with respect to the sphere. This puts you outside the cage and the lion inside. Boas published the paper under the pseudonym of H. Pétard. He later explained that Pétard's full initials are H. W. O., which stand for "Hoist With Own." Several mathematicians were so amused by this paper that they later published articles, all included in this anthology, giving additional methods for lion trapping.

This article first appeared in *American Scientist* (March/April 1996).

271

The paper about snowfalls and elephants is an introduction to how certain statistical problems can be solved by careful sampling. Boas first takes up the way to calculate the expected number of heavy snowfalls in a particular place or during a person's lifetime, then shows how similar techniques apply to finding an extremely heavy elephant in a herd without having to weigh them all and, finally, how to determine the number of soft-drink bottles that will not explode when the temperature soars.

Two other nontechnical papers merit special mention. "Distribution of Digits in Integers" reveals some startling consequences of the fact that almost all real numbers are "normal" in the sense that every digit or finite run of digits occurs equally often in the digital expansion of that number in any base notation. Imagine that the *Encyclopedia Britannica* is coded by numbers that form one monstrous integer. If such a real number as π, e or $\sqrt{2}$ is indeed normal, as suspected though not proved, then the *Britannica*, in its numerical form, will occur an infinite number of times in the decimal expansion of the number.

"Traveling Surprises" applies the mean value theorem to a car that goes from here to there with an average speed of 50 m.p.h. Boas proves there must be an instant at which the car's speed is precisely 50 m.p.h., and that there is a short interval during which the average speed is 50 m.p.h. Even more surprising, if the car travels longer than an hour, there must be one hour during which the car goes exactly 50 miles. The paper includes a graph with an arrow pointing to a spot on a space–time curve. The arrow is labeled "Aha!"

Many of the book's anecdotes are so technical that only mathematicians will find them funny. Some even require a knowledge of German. I particularly enjoyed those anecdotes that convey Boas's love of word play. That "ghoti" spells "fish" is so well known that James Joyce mentions it in *Finnegans Wake*. (You pronounce "gh" as in laugh, "o" as in women, and "ti" as in nation.) Not well known is Boas' discovery that "ghghgh" spells "puff." Pronounce "gh" as in hiccough, Edinburgh and laugh.

Norbert Wiener once said he would like to write a story about a girl from Walla Walla who went to Pago Pago to dance the hula hula. This inspired Boas to write

> Lulu from Walla Walla was a devotee of dance:
>   She did a wicked can-can in a tutu sent from France.
> She said, "I'm going gaga; in toto, life's a bore,"
>   So she went to Bora Bora and to Pago Pago's shore,

Where she studied hula hula and tried for an A.A.,
But her work was only so-so, and they wouldn't let her stay.
Now she's gone to Baden Baden, to a go-go cabaret;
Billed as Lulu in the muu muu, she's performing every day.
When she lets the muu muu drop, the old folks drop their teeth
When they see a lava-lava is all she has beneath.

In Princeton's Fine Hall, Boas recalls, someone once posted a "Scale of Obviousness":

If Wedderburn says it's obvious, everybody in the room has
seen it ten minutes ago.
If Bohnenblust says it's obvious, it's obvious.
If Bochner says it's obvious, you can figure it out in half
an hour.
If von Neumann says it's obvious, you can prove it in three
months if you're a genius.
If Lefschetz says it's obvious, it's wrong.

Boas was puzzled by a Japanese paper that kept referring to "stricken mass distributions." Unable to figure out what this meant, he wrote to the journal's editor. It turned out that a referee's report had told the author, "The term 'generalized mass distribution' is no longer used. The word 'generalized' should be stricken."

There are simple ways to determine whether a number is divisible by each digit except 7. Boas recalls a letter proposing a way to test for 7. "Because an integer is divisible by 10 if and only if its last digit in base 10 is 0, then to test its divisibility by 7 you need only write the number in base seven and look at the final digit."

Boas remembers Grünwald's theorem in algebraic number theory. "It had a very long proof," he writes, "but after a while Whiples found a shorter proof. Later, somebody found a counterexample." Papers often have uninformative titles. Boas cites the worst example he ever came across. It was titled "On a Certain Theorem." He recalls a review of another paper that said, "This paper contains two theorems. The first is due to the referee and the second is wrong."

The short story by Boas that I enjoyed most is about Valentina, a mathematical whiz who creates a computer program that plays perfect chess, winning every game. Rather than destroy the game, she locks it in a safety deposit vault for 100 years, hoping that by then the chess world will be able to accept it. I have, of course, touched only a few aspects of this delightful anthology that conveys so well the depth and wit of a remarkable man. The book closes with a bibliography of more than 280 references to books and articles by Boas, including his translations of Russian books, but excluding hundreds of his book reviews.

Chapter 36
Two Books on Infinity

Since ancient times, mathematicians, scientists, philosophers and theologians have struggled with the dark mysteries of infinity. Our universe seems to be of finite size, expanding at an unknown rate ever since the Big Bang, but is there an infinite space into which it is expanding? Time also seems to have started with the bang, but was there some sort of super time before then? And if our universe has enough matter to halt the expansion and send the cosmos backward toward a Big Crunch, will time continue to march on?

For Immanuel Kant, the infinities of time and space were contradictory "antinomies." It is difficult to imagine space extending forever, but it is just as hard to suppose it stops, because that will lead you to ask: "What's outside the boundary?" It is hard to imagine time without beginning and end but equally hard to suppose otherwise.

Cosmologists today are speculating, as did the 16th century Italian philosopher Giordano Bruno and others before him, that there may be an infinity of isolated universes, each with its own set of laws. Only a tiny subset of these universes allows life to arise. As someone has said, a universe is something that happens now and then. Time, someone else said, is what keeps everything from happening at once.

Infinity also extends inward. Is there an ultimate particle, or does matter have infinite levels of structure like a set of Russian dolls? Science fiction writers have long played with the concept of infinite universes. H. G. Wells wrote a story in which our cosmos is an atom in a ring worn by a gigantic hand. Other writers have turned our atoms into other universes. Cyberpunk writer Rudy Rucker, in his novel *Spacetime Donuts*, introduced a wild circularity. He imagines the paradox of a person getting smaller and smaller, visiting ever tinier universes, until he finds himself back in the universe where he started without having to change his size.

This review appeared in The *Los Angeles Times Book Review* (May 18, 1997).

Mathematicians speak of parallel lines meeting at infinity and of infinite series of fractions whose partial sums converge at infinity on a well-defined limit. Calculus rests on the summation of infinitely small quantities that German philosopher and mathematician Gotfried Wilhelm von Leibniz called infinitesimals. They were much derided by Bishop George Berkeley in the early 18th century and considered meaningless by most mathematicians until they became respectable again in what is known as nonstandard analysis.

A regular polygon turns into a circle when it has an infinity of sides. The digits of irrational numbers like π go on endlessly without repeating a particular sequence of numbers. A revolution in mathematics occurred when German mathematician George Cantor, in papers published between 1874 and 1884, found a way to define a hierarchy of infinities.

The lowest infinity represents the number of integers, including integral fractions. The next higher one counts the real numbers. The next counts all the curves that can be drawn on a postage stamp. Cantor's transfinite numbers, which he called alephs, are themselves infinite in number.

Theology, too, has its infinities. Does God have infinite attributes? Is God subject to time and change or does God exist in eternity, outside space and time altogether? After we die, do we suffer eternally in hell or live forever in paradise?

A number of interesting books have been written about infinity, including a nonfiction book by Rucker, *Infinity and the Mind*. The most recent are Richard Morns' *Achilles in the Quantum Universe* (Henry Holt, 1997), J. V. Field's *The Invention of Infinity* (Oxford University Press, 1997), and Eli Maor's *To Infinity and Beyond* (Springer Verlag, 1991).

Morris, author of several books about science, has done his usual excellent job of ranging over his topic in lucid, entertaining prose. He covers all the important questions and more. The Achilles of his book's title is the Greek warrior in one of Zeno's famous paradoxes of motion. Achilles is unable to catch a tortoise because when he reaches the spot where the turtle started, the reptile has crawled ahead a finite distance. When he goes that distance, the turtle has moved ahead by a smaller distance. The distances the turtle moves ahead keep shrinking but never vanish, so how does Achilles manage to overtake the tortoise? The two runners model dimensionless points that move along a straight line. If Achilles were to pause a minute after running each segment, he would never reach the turtle. As it is, if both go at a steady rate, the time it takes Achilles to traverse each segment also gets smaller, converging on zero, allowing him to reach the tortoise in a finite time.

Morris covers all the infinities that plague cosmology, particle physics, relativity theory and quantum mechanics. The infinities that arise in black holes are especially troublesome. In a black hole, the volume of matter becomes zero and its density goes to infinity. No one knows just what happens after that. Does the matter explode, as some cosmologists suggest, from a white hole in another universe?

Morris does his best to explain a bizarre model of the universe recently proposed by Stephen Hawking. In ordinary time, our universe has a beginning and end, but in "imaginary time" (based on imaginary numbers), the universe is infinite in both directions. As far as I can tell, no one except Hawking has taken his model seriously.

It is a credit to Morris that in writing about infinity, he draws upon understandable and entertaining analogies that have historically been used to describe infinity. For example, in explaining Cantor's alephs, Morris introduces the notorious aleph hotel, which has an infinity of rooms numbered 1, 2, 3,... Every room is occupied. Can the manager accommodate the arrival of an infinity of guests? Easy. He simply moves each occupant to a room with a number twice his or her previous number. This opens up an infinity of rooms, namely all those with odd numbers. As Morris informs us, this paradox was recognized by Galileo, but it took Cantor to clarify the mathematics involved.

Morris concludes his book with the concept that "the difficulties encountered by modern physicists show us that the infinite is still as much a mystery as it was in the time of Zeno." Both atheists and theists, he rightly maintains, can accept the now fashionable notion of a plurality of universes, perhaps even an infinity of them, exploding here and there, now and then, in some kind of super time and space. Meditating on such possibilities can arouse in one a sense of awe so intense that if it persisted more than a minute or two, one could go mad. Morris' final sentence is a memorable quote from Pascal, as pious a theist as he was a great mathematician: "The eternal silence of these infinite spaces frightens me."

Field's *The Invention of Infinity*, in spite of its title, has very little in common with Morris' book. A research fellow in art history at Birkbeck College, University of London, Field has written books about Kepler's cosmology, the geometry of Girard Desargues and *Sci-*

ence in Art, beautiful work featuring paintings in London's National Gallery that portray aspects of science and technology. *The Invention of Infinity* is an equally handsome volume reproducing hundreds of striking pictures, alas none in color, that relate to mathematical ideas of their time.

Although Field makes side excursions into the history of geometry and algebra, the primary focus of his book is the development of projective geometry and its application to perspective in painting. It is here that infinity comes into play as the "distance point" on the horizon at which parallel lines meet at an infinite distance from the viewer. Field's book is essentially a detailed, erudite treatise on the collaboration of eminent Renaissance mathematicians with artists who had mastered the art of perspective.

Field reproduces and discusses numerous paintings by Italian artists, many of whom also wrote treatises on perspective. Mathematicians who contributed to perspective and the concept of infinity include Kepler, Pascal, Descartes and, above all, Desargues, the French mathematician who wrote the first great work on projective geometry in the 17th century.

Field does not neglect the conic section curves that result when a right circular cone is sliced by a plane. If the plane is parallel to the cone's base, the section is a circle. Tilt the cone ever so slightly and the circle's center splits into two foci to create an ellipse, the shape a circle has when seen in perspective. As the plane assumes steeper angles, one focus moves father from the other until the plane is parallel with the cone's side. At that point, the focus traveled to infinity and the ellipses have become a parabola with arms that are parallel when they are infinitely far from the other focus. As the cutting plane tips even more—until it is perpendicular to the cone's base—the cross-sections turn into parabolas. Their arms meet their asymptotes at infinity.

In his last chapter, Field also refers to Pascal's remark about his fear of infinite space. Field seems to think that Pascal is expressing not his own fears but the fears of an atheist. I believe this is a misreading of Pascal. It is Pascal's own terror, the terror of a devout believer in God, that Pascal is describing. Here is a longer quotation from Pascal's "Pensees," a work saturated with wonder about the awesome mysteries of space and time:

> When I consider the short duration of my life, swallowed
> up in the eternity before and after, the small space which
> I fill, or even can see, engulfed in the infinite immensity of

spaces whereof I know nothing, and which know nothing of me, I am terrified, and wonder that I am here rather than there, for there is no reason why here rather than there, or now rather than then. Who has set me here? By whose order and design have this place and time been destined for me?

Chapter 37
The Universe and the Teacup

The Universe and the Teacup: The Mathematics of Truth
and Beauty. by K. C. Cole. Harcourt Brace, 1998.

On an acknowledgments page, science writer K.C. Cole says she was
surprised when her editor told her she had written a book about mathematics. She had intended to write a book that surveyed trends in
contemporary science, and she has, delightfully, but math is the common thread that binds it all together. What do the largest galaxy and
the smallest teacup have in common? Answer: mathematical structure. Cole's book is a loving paean to the awesome power and beauty
of mathematics.

Cole's early chapter on giant numbers stresses how hard it is to
grasp their applications. Our galaxy contains some 200 billion stars,
and there are more than 200 billion galaxies. "Number numbness" sets
in when you contemplate the vastness of space and time. A geologist,
after chalking on the blackboard a line that runs from zero to a trillion,
asks where to put the point that indicates a billion. Amazingly, it is
extremely close to zero! "Compared to a trillion," Cole comments, "a
billion is peanuts."

Few realize how rapidly numbers grow by a simple process of dou-
bling. Put two grains of wheat on the first square of a chessboard,
four on the next, then eight and so on. You can't get very far. After
only 10 steps, a square requires 1024 grains. The last square needs
more grains than have ever been produced on Earth. Big numbers.
Cole tells us, "creep up on us unawares." Disaster looms when pop-
ulation grows exponentially on a planet with a surface as finite as a
chessboard.

Widespread failure to understand simple probabilities receives careful treatment. We worry more about alar in apples, Cole writes, than

This review appeared in The Los Angeles Times Book Review (February 1, 1998).

the greater threat of cigarette smoke. A woman may avoid fatty meats, yet not mind sleeping with strangers. Everywhere there is what Cole calls "skewed perception of risks."

In a chapter on measurements, Cole introduces the central mystery of quantum mechanics. How can an electron have no definite properties, such as position or spin, but acquire precise attributes as soon as they are measured? She likens this to a spinning coin, which is neither heads nor tails until it falls flat. Difficulties in measuring intelligence are also considered.

There are also severe limits on how large certain things can be. No teacup can be as big as Jupiter because gravity would pull it into a sphere. Gravity would fracture the thighs of a person 60 feet tall. A flea can jump high in relation to its size, but the jump's height, approximately one meter, is about the same as what an average person can achieve. Cole calls it an "interesting invariant" that most animals can't leap higher than that.

A thousand other fascinating facts and shrewd observations crowd into Cole's lyrical praise of science, from the wonders of galaxies to the invisible creatures that live on our eyelashes. She is good at emphasizing how unpredictable properties continually emerge as the universe evolves from primeval simplicity to the enormous complexity of a human brain. What could be more unlike a gas than water? Yet when two gases, hydrogen and oxygen, combine, the more complex properties of water emerge. Stars and rocks result from even more complex combinations of molecules. At some time in the distant past, life sprang into existence from the complexity of a self-replicating molecule. As it evolved from simple one-cell forms, the magic of increasing complexity produced trees, butterflies, dinosaurs, frogs and elephants.

You and I somehow emerged, after billions of years, from the quantum fields that predated the big bang. Biology is not applied physics, Cole reminds us, and psychology is not applied biology. Knowledge of sounds reveals nothing about a Mozart symphony. "You can learn everything there is to know about the atoms that make up a cat," Cole writes, "and that still will not tell you whether it will scratch the furniture or sleep on your head."

Slow increments of change lead to "tipping points," where a qualitative change occurs suddenly. Water cools gradually, then, presto, it turns into ice. Tipping points are all over the human scene: an unanticipated drop in crime, a plunge of the Dow, an angry man's murder of his wife. Chaos theory studies systems that start out orderly, then, by altering in a completely deterministic way, suddenly cross a threshold to become unpredictably random.

"The Signal in a Haystack" is Cole's title for a chapter on how science does its best to filter significant facts from vast amounts of irrelevant information that often contaminate research results. Astronomical data are still not free enough from cosmic noise to determine how fast the universe is expanding or exactly how old it is. Are there really black holes at the centers of galaxies? Does star wobbling indicate planets? Are wiggly forms on a meteorite fossils from Mars? "The sky is like a cocktail party," Cole writes, "with too many conversations going on at once." Scientists listen to the babble in the hope of hearing what the universe is trying to tell them. As Cole writes: "That's exactly how the search for truth is supposed to work. You see something, and then you try everything you can think of to make it go away; you turn it upside down and inside out, and push on it from every possible angle. If it's still there, maybe you've got something."

The Universe and the Teacup contains informative chapters on the inevitable flaws in every voting system and on how things can be divided fairly. Fair division is a piece of cake when only two people want half of a piece of cake. One person cuts; the other chooses. The trick has endless applications to social and political conflicts, such as who gets what after a divorce, assuming each party wants to be "envy free." If more than three players are involved, the task of fair division becomes more difficult. Ingenious solutions by various experts are skillfully outlined.

The conflict between altruism and egoism also gets Cole's attention. Does self-sacrifice have survival value for an evolving animal species? Can bitter conflicts between business firms or nations be settled amicably by applying game theory? The famous tit-for-tat strategy is explained. This involves competition between two "players," such as the effort of one nation to establish free trade with another nation or to obtain mutual disarmament. The tit-for-tat strategy is to make the first move in the desired direction. After that, do whatever your adversary does. Eventually it dawns on both players that in the long run, cooperation maximizes self-interest.

Near her book's close, Cole tackles the question of whether scientific truth is always fallible in contrast to the certainty of math and logic. Kurt Godel's famous undecidability theorem (it states that every formal system of mathematics that includes arithmetic contains true statements that cannot be proved true within the system) is considered and so is the current fad of fuzzy logic. The book's final chapter is a tribute to Emmy Noether, an eminent German mathematician. She proved that behind every invariant in physics lurks a symmetry that can be defined by an algebraic structure called a group, an insight essential to relativity theory and particle physics. Immediately after the big bang, the universe was inconceivably hot and perfectly symmetrical. As it cooled, various symmetries were shattered to fashion the cool and broken universe we know and love.

I have touched on only a tiny fraction of the topics covered in this dense and passionate book. You will put it down sharing Cole's awe and wonder at the vastness and intricacy of what G. K. Chesterton once said an atheist must view as the most exquisite mechanism ever constructed by nobody.

Chapter 38
A New Result on Perfect Magic Squares

Beauty is the first test:
There is no permanent place in the world
for ugly mathematics.
—G.H. Hardy, *A Mathematician's Apology.*

Dame Kathleen Ollerenshaw, one of England's national treasures, has solved a long-standing, extremely difficult problem involving the construction and enumeration of a certain type of magic square. But first, some background.

For many centuries mathematicians, especially those concerned with combinatorics, have been challenged by magic squares. These are arrangements of n^2 distinct integers in an $n \times n$ array such that each row, column, and main diagonal has the same sum. The sum is called the *magic constant*, and n is the square's *order*. Traditional magic squares, known as "normal," are made with consecutive integers starting with 0 or 1. If a magic square starts with 0 it can, of course, be changed to start with 1 simply by adding 1 to each cell.

No order-2 square is possible. The order-3, shown in Figure 1, barely exists. Why? Because there are just eight different triplets of distinct digits from 1 through 9 that add to 15, the square's constant. (If a square starts with 1, the constant is half the sum of n and n^3.) Each triplet appears as one of the square's eight straight lines of three numbers. The pattern is unique except for trivial rotations and mirror reflections, which are never considered different.

This, little gem of combinatorial number theory was called the *lo shu* in ancient China. Legend has it that in the 23rd century B.C.

A cut version of this review appeared in *Nature* (Vol. 395, September 17, 1998).

Figure 1. The *lo shu*, the only magic square of order 3.

a mythical King Yu saw the pattern on the back of a sacred turtle in the River Lo. Modern historians, however, find no evidence that the pattern was known before the fourth or fifth century B.C. At any rate, the name means Lo River Writing. The Chinese identify it with their familiar yin-yang circle. The even digits, here shown shaded, are linked to the dark yin. The Greek cross of odd digits is linked to the light yang. For centuries the *lo shu* has been used as a charm on jewelry and other objects. Today, large passenger ships often use the *lo shu* as a shuffleboard pattern.

The number of order-4 magic squares jumps to 880. Among them is a famous subset of 48 squares called pandiagonal. (In the past they have been called *diabolic* and *nasik squares*.) A pandiagonal square with a constant of 30 is shown in Figure 2.

All pandiagonal squares of order 4 have three amazing properties:

1. Each "broken diagonal" also adds to 30. (Cells 0, 3, 4, 15, 12 and 7, 13, 8, 2, and so on, are examples of broken diagonals.)

0	13	6	11
7	10	1	12
9	4	15	2
14	3	8	5

Figure 2. An order-4 most-perfect square. The magic constant is 30.

Imagine an endless array of this square, or any other order-4 pandiagonal, placed side by side in all directions to make a wallpaper pattern. Every 4 × 4 square drawn on this pattern will be a pandiagonal magic square. Every straight line of numbers, orthogonally or diagonally, will add to 30. Similar properties hold for pandiagonal squares of all higher orders. A 5 × 5, for example, will form a wallpaper pattern on which every 5 × 5 square is a pandiagonal.

2. Every 2 × 2 square on the wallpaper also adds to 30. Imagine that the square wraps around horizontally and vertically to form the surface of a torus. Any 2 × 2 square on this torus will have a sum of 30.

3. Along every diagonal of the wallpaper, any two cells separated by one cell add to 15.

If rotations and reflections are included, there are $8 \times 48 = 384$ order-4 pandiagonal squares. If a square's order is even, it must be a multiple of 4. If its order is odd, it can be any number.

A pandiagonal square is "most-perfect" if, like the order-4 pandiagonals, all its rows, columns, and diagonals (main and broken) add to the magic constant, and all the 2 × 2 squares on its wallpaper pattern have the same sum. If the sguare starts with 0, its constant is $1/2(n^3 + n) - n$, and each 2 × 2 square adds to $2(n^2 - 1)$. moreover, along every diagonal (main and broken), numbers that are $n/2$ cells apart add to $n^2 - 1$.

Although all order-4 pandiagonal squares have been known to be perfect for three centuries, very little has been known about most perfect squares of higher orders. There was no known method for constructing all of them or determining the number of such squares for any given order. These were the two unanswered questions finally settled by Kathleen Ollerenshaw. Her solution is the main topic of a wonderful little book titled *Most-Perfect Pandiagonal Magic Squares: Their Construction and Enumeration*. Written with David Brée, and with a foreword by the cosmologist Sir Hermann Bondi, the book is scheduled for publication by the Institute of Mathematics and Its Application, at South-end-on Sea, Essex.

Because all most-perfect squares are pandiagonal, their even orders must be a multiple of 4. However, unlike the pandiagonals, they do not exist with odd orders. Figure 3 shows a most-perfect square of order 8. As a wallpaper pattern, every diagonal line of eight cells will

0	62	2	60	11	53	9	55
15	49	13	51	4	58	6	56
16	46	18	44	27	37	25	39
31	33	29	35	20	42	22	40
52	10	54	8	63	1	61	3
59	5	57	7	48	14	50	12
36	26	38	24	47	17	45	19
43	21	41	23	32	30	34	28

Figure 3. A most-perfect square of order 8. The magic constant is 252.

add to the square's magic constant of 252. This is the same as saying that all its broken diagonals add to 252. Every 2×2 square on the pattern adds to $2(8^2 - 1) = 126$, and every pair of numbers that are $8/2 = 4$ cells apart along a diagonal add to $8^2 - 1 = 63$.

Because the orders of most-perfect magic squares increase by leaps of 4, the number of essentially different most-perfect squares increases very rapidly as n increases. The authors have determined for the first time a method for constructing all most-perfect squares of any order, and a complicated formula that counts the number of such squares for any order. The number of essentially different order-8 squares is 368,640. The number of order-12 squares is 2.22953×10^{10}. When you reach order 36 the number is 2.76754×10^{44}. Assuming that the Big Bang occurred 20 million years ago, the authors estimate that the number of order-36 most-perfect squares is more than a thousand times the number of pico-pico-seconds since the Bang. (A pico is one trillionth of a number.)

The author's elegant solution of one of the most frustrating of unsolved problems in magic square theory was an achievement that would have been remarkable for a mathematician of any age. In Dame

Kathleen's case this was even more remarkable because she was 85 when she and Brée finally proved the conjectures she had earlier made intuitively. Most mathematicians do their most creative work when quite young. Dame Kathleen is a striking exception.

Kathleen Ollerenshaw is surely one of the most amazing women in England. Born in Manchester in 1912, she obtained her doctorate in mathematics at Somerville College, Oxford. It would take many inches just to list her many awards and the positions she has held in her long and colorful life. They include president of the Institute of Mathematics, deputy president of the University of Manchester Institute of Technology, and the Lord Mayor (1975–76) of Manchester. Dave Brée (born 1939), coauthor of the book, is a professor of artificial intelligence at the University of Manchester.

The author's marvelous solution to the problem of constructing and counting all most-perfect magic squares is too technical to go into here, but her book explains it all as clearly as possible. At the back of the book, in a section headed "A Personal Perspective," Dame Kathleen concludes:

> The manner in which each successive application of the properties of binomial coefficients that characterize the Pascal triangle led to the solution will always remain one of the most magical mathematical revelations that I have been fortunate enough to experience. That this should be afforded to someone who had, with a few exceptions, been out of active mathematics research for over forty years will, I hope, encourage others. The delight of discovery is not a privilege reserved solely for the young.

References

[1] Andrews, W. S. *Magic Squares and Cubes* (Dover, New York, 1960).

[2] Benson, William A., and Jacoby, Oswald. *New Recreations with Magic Squares* (Dover, New York, 1976).

[3] Hirayama, Akira, and Abe, Gakuho. *Researches in Magic Squares* (Osaka Kyoikutusho Co., Osaka, Japan, 1983.)

[4] Ollerenshaw, Dame Kathleen, and Bondi, Sir Herman. Magic squares of Order Four. *Philosophical Transactions of the Royal Society*, 306, 443–552 (1982).

[5] Rosser, B., and Walker, R.J. The algebraic theory of diabolic magic squares. *Duke Mathematical Journal*, 5, 705–728 (1939).

[6] Stewart, Ian. Most-Perfect Magic Squares. *Scientific American*, November 1999. A good summary of Dame Ollerenshaw's book.

Chapter 39
The Number Devil

No book about mathematics, written for young children, could less resemble a textbook than *The Number Devil* (Metropolitan Books, 1998). The author, Hans Magnus Enzensberger, who lives in Munich, is a writer and scholar, but not a mathematician. This may explain how he manages to introduce number theory in such an entertaining way that his book became a best seller in Germany. Translated by Michael Heim, and amusingly illustrated by Rotraut Susanne Berger, this is just the book to give to an intelligent child who falls asleep in mathematics classes.

Enzensberger imagines a twelve-year-old Robert who hates math because his teacher Mr. Bockel—the name means an obstinate goat in German—is such a stupid and dull teacher. One night, after some unpleasant nightmares, Robert dreams about a friendly imp with shining eyes who calls himself a number devil. In this and eleven later dreams, the devil explains elementary number theory in such a refreshing way that Robert, instead of being bored, is instantly intrigued.

The devil's instructions during the first dream couldn't be simpler. Every integer, he explains, is reached by adding ones. Because this process can go on forever, there must be an infinity of integers and no such thing as a largest number. Similarly, $\frac{1}{1}, \frac{1}{11}, \frac{1}{111}, \ldots$ generates a infinity of smaller and smaller fractions, never reaching a smallest fraction.

When 11 is multiplied by 11 you get 121, a number palindrome that is the same backward, like such words as MADAM and ROTATOR. and such sentences as "Straw? No, too stupid a fad. I put soot on warts." Number palindromes also result from $111 \times 111 = 12321$, and $1111 \times 1111 = 1234321$. Does this continue to produce palindromic products as the number of ones increases? No. Robert correctly guesses

---

A cut version of this review appeared in The *Los Angeles Times Book Review* (November 8, 1998).

that the pattern fails beyond ten ones. Moral: You can't trust a generalization until it is proved.

Things are slightly more complicated in Robert's second dream. The number devil convinces him that Roman numerals were such a clumsy notation that it held mathematics back for centuries. Robert learns the value of the decimal system, with digits ordered from right to left, the last digit indicating a multiple of one, the preceding digit a multiple of ten, the next digit a multiple of one hundred, and so on, with zero serving as what mathematicians call a "place holder."

In his third dream Robert learns the importance of prime numbers, numbers evenly divisible only by themselves and one, and how to find them by a sieving method. He is told about a famous theorem (known as Goldbach's conjecture), still not proved, that every even numbers greater than two is the sum of at least one pair of primes. (For example, 1998 is the sum of primes 1993 and 5.)

The devil reveals a curious fraction in Robert's fourth dream. One divided by seven produces an endless decimal fraction 0.142857142857142857.... Its repeating pattern 142857 has the surprising property that when it is multiplied by any digit from one through six the quotient has the same digits in the same cyclic order. Irrational fractions, which have no repeating pattern in their decimal expansion, come next. The devil proves that the diagonal of a square of side one has a length equal to the square root of two, an irrational number that begins 1.414213....

The devil has his own whimsical terminology. Irrational numbers are called "unreasonable numbers." Roots are called "rootabagas." Primes are "prima donnas." At the back of the book there is a handy list of the devil's terms translated into the terminology of modern mathematics.

Triangular and square numbers appear in Robert's fifth dream. Triangular numbers (1, 3, 6, 10, 15, ...) are integers that can be modeled by dots in triangular arrays, like the fifteen pool balls or the ten bowling pins at the start of a game. Square numbers (1, 4, 9, 16, ...) are modeled by dots in square arrays. Figure 1 shows the devil's elegant "look–see" proof that every partial sum of the series of odd numbers is a square number.

The fascinating properties of Fibonacci numbers (1, 1, 2, 3, 5, 8, 13, ...) are the topic of the sixth dream. Each number is the sum of its two predecessors. They are followed in the next dream by the wonders of Pascal's famous number triangle. Dream 8 concerns permutations

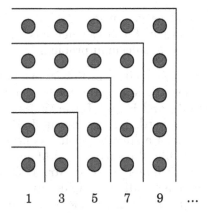

1   3   5   7   9   ...

Figure 1. A look–see proof.

and factorial numbers such as 6! where the exclamation mark tells you to multiply $1 \times 2 \times 3 \times 4 \times 5 \times 6$. It numbers the different ways six students can sit in a row.

Here are four of the topics covered in the last four dreams:

1. The proof (first noticed, by the way, by Galileo) that there are as many odd numbers as there are counting numbers. You simply pair them like so:

$$
\begin{array}{ccccc}
1 & 2 & 3 & 4 & 5 \quad \ldots \\
\downarrow & \downarrow & \downarrow & \downarrow & \downarrow \quad \ldots \\
1 & 3 & 5 & 7 & 9 \quad \ldots
\end{array}
$$

The number of numbers in both sets is a "transfinite" number that mathematicians know as aleph null, the smallest of an infinite set of transfinite numbers. Although the odd numbers form an infinite set, and the counting numbers form another infinite set, the two infinities have a sum that is the same transfinite number.

2. The beautiful properties of ϕ, the golden ratio 1.618... It is a famous irrational number that is the limit approached by the ratios of adjacent Fibonacci numbers.

3. The five Platonic solids, the only convex polyhedrons with faces that are regular polygons. The devil teaches Robert how to make them by cutting and folding paper.

4. The difficult and as yet unsolved task of finding the shortest
route that visits each of n cities when n is large. Mathematicians
call it the "traveling salesman problem."

In the final dream the devil takes Robert to a party where he meets
some eminent mathematicians of the past. They include Lord Rus-
tle (Bertrand Russell), and Happy Little (Felix Klein). Little shows
Robert a strange glass bottle he has invented. Its edgeless surface is
closed like the surface of a sphere, yet it has no inside or outside. It
is a one-sided surface like the surface of a Möbius band.

The book ends with Robert back in Mr. Bockel's class, where he
surprises the teacher by finding a rapid way to solve a problem. If
Mr. Bockel gives a pretzel to one student, two to another, three to a
third student, and so on for 38 students, how many pretzels does he
hand out?

Adults who know little about math will find this book as enlight-
ening as will younger readers. It closes with a valuable index that
lists in standard terminology all of the number devil's topics.

Chapter 40
Probability 1

If one is on a hunt, it is better
not to assume at the start that there is no game, or,
you won't get what little there is.
—William James (in a 1902 letter)

The debate over whether there is intelligent life elsewhere in the universe is an ancient one. Greek and Roman atomists (Leucippus, Democritus, Epicurus, Lucretius) all defended a plurality of worlds. Plato and Aristotle thought otherwise. Medieval Christian thinkers, from Augustine to Aquinas, followed Aristotle and the Bible by confining mortal life to Earth. Pluralism revived during the Renaissance, notably in the writings of the German cardinal Nicholas of Cusa and the Italian ex-monk Giordano Bruno. (It is often said that Bruno was burned at the stake for his belief in many worlds, but it was mainly for other heresies.)

The Copernican revolution, which removed Earth from the center of the cosmos, surely played a role in the intense upwelling of pluralism in post-Renaissance centuries. Kant, Newton, Pope, Voltaire, Paine, Emerson, together with hundreds of writers and scientists, eagerly embraced the notion that intelligent life was everywhere, most likely on Mars, perhaps even on the moon. The leaders of America's two greatest Adventist movements—Joseph Smith, who started Mormonism, and Ellen White, founder and prophetess of the Seventh-day Adventists—each defended a plurality of inhabited worlds. White, like the Swedish mystic Emanuel Swedenborg, even had visions of humanlike beings on other planets. (Two eminent English exceptions to this belief were the naturalist Alfred Russel Wallace and philosopher of science William Whewell. Each argued strenuously against the possibility of extraterrestrial life.)

This review appeared in The *Los Angeles Times Book Review* (November 29, 1998).

A widespread view among today's scientists is that the universe probably teems with life. There are billions of galaxies, each with billions of stars, and in recent years evidence has been increasing for a plenitude of other solar systems. So far only planets huge enough to cause detectable wobblings of their mother suns have been found, but there is every reason to believe that most suns, perhaps all suns, have planets of all sizes orbiting them.

In recent years, Carl Sagan was the most vocal scientist to trumpet a belief in the plurality of inhabited planets and to urge continual funding of searches for radio messages from extraterrestrial intelligence. Frank Tipler, a physicist at Tulane University, is Sagan's chief detractor. Tipler thinks the probability of extraterrestrials is zero and that listening for their signals is a big waste of time and money. Less aggressive skeptics include physicist John Wheeler, biologists Ernst Mayr and Jacques Monod, mathematician John Casti, astronomer Martin Rees and a raft of other scientists.

Amir D. Aczel, a statistician at Bentley College in Waltham, Mass., whose previous book, *Fermat's Last Theorem*, won a 1996 Los Angeles Times Book Prize, is even more certain than Sagan that sentient life is not limited to Earth. In his new book, *Probability 1* (Harcourt Brace, 1998) as its title suggests, he maintains that the probability of beings on at least one other planet is so close to 1 as to be indistinguishable from certainty. Although his book covers the same ground as earlier books by others, there are two reasons for recommending it to any person interested in the debate: It is clearly and gracefully written, and it is up to date in its astronomical data.

In 1950, Aczel reminds us, the Italian physicist Enrico Fermi asked a famous question now called Fermi's paradox. The universe is so vast and so old, Fermi said, that if intelligent life is out there, we can expect many civilizations to have a technology far in advance of our own. Long ago they should have visited Earth. "Where are they?" Fermi asked. Because we have not seen them or heard from them, they probably don't exist. Like almost all scientists, Fermi dismissed UFO mania as popular superstition.

Many scientists refused to buy Fermi's argument. There are too many reasons, they said, why extraterrestrials would find it difficult to cover the vast distances between solar systems. Moreover, why would they select Earth from billions of other planets in our galaxy?

Aczel retells the story of how, in the late 1950s, physicist Philip Morrison, astronomer Frank Drake and others concluded that extraterrestrials might be sending messages using a certain frequency of radio waves. Drake's Project Ozma, named after the Princess of Oz in L. Frank Baum's popular Oz series, was the first attempt to listen for such messages using a radio telescope at Green Bank, W. Va. Others soon joined in the search. In 1993, a skeptical Congress stopped funding these searches. Drake, Aczel tells us, has continued his work with private funding from corporations. The most efficient search, SERENDIP III, is being conducted by Stuart Bowyer, using the world's largest radio telescope in Arecibo, Puerto Rico. Since 1992, it has examined 500 trillion signals without detecting a message.

In subsequent chapters, Aczel skillfully covers the recent discoveries of some dozen large planets circling nearby suns, theories about how life arose on Earth and the remote possibility that life could be based on elements other than carbon, as found on Earth. Aczel considers silicon and sulfur, concluding that neither has anywhere near the flexibility of carbon in forming large protein molecules so essential for the specialized tasks they perform in Earth's life forms. If there is extraterrestrial life, Aczel concludes, it is probably carbon-based. Another chapter deals with "panspermia," the conjecture that life on Earth began with a seeding of organic material carried here by comets or meteors. This old conjecture has been revived by astronomer Fred Hoyle and biologist Francis Crick.

The famous Mars meteorite discovered in Antarctica in 1996, containing what could be fossil bacteria, is evaluated by Aczel and found to be inconclusive. A chapter is devoted to the influence on Earth's evolving life by the impacts of large meteors, asteroids and comets. Aczel defends the view that intelligence is not just an accidental product of evolution, as Stephen Jay Gould and others believe, but the inexorable outcome of a tendency for life, once it starts, to evolve creatures with ever superior brains.

The chapter "Does God Play Dice?" introduces the random aspects of quantum mechanics, which Einstein found so distasteful, along with the determinism of chaos theory and Benoit Mandelbrot's fractals. These topics lead into deep questions about why evolution on Earth produced creatures capable of creating art, music, poetry, mathematics, science and philosophy, none of which seems to have much Darwinian survival value. Are they mere byproducts of a brain designed by nature only to increase efficiency in hunting, fishing, fighting and growing food? Or do they have survival value in some deeper sense not yet understood?

Another chapter concerns what statisticians call the "inspection paradox." Assume, writes Aczel, that a bus arrives at a bus stop on the average of every 10 minutes. You go there at random moments. One might suppose that the average waiting time would be five minutes, but this is not the case. It's a trifle longer than five minutes. Why? If the bus arrived at the stop at exactly 10-minute intervals, the average wait would indeed be five minutes. The bus, however, arrives at 10-minute intervals only on the average. If you go to the stop at a random time, the probability is higher that you will arrive at the stop within a long time interval between bus arrivals than during a short interval. Your average wait, therefore, will be longer than five minutes.

To further clarify the paradox, Aczel gives another example. Assume that a flashlight battery has an average life of $3\frac{1}{2}$ hours. Each time you pick up the flashlight you are selecting a random point of time between the random durations between battery use. Like the bus rider, you are more likely to pick up the flashlight between long time intervals than short ones. Result: Your battery will last a trifle longer than its average life.

Aczel applies these examples to Fermi's question. "If your arrival on Earth is viewed as a random event, you are more likely to land on a longer-lived planet than a shorter-lived one in the same way that a dart is more likely to land on a wider section of the dart-board than on a narrower one." Our sun is known to have a longer-than-average lifetime. "It is very likely that, as galactic civilizations go, we are on the above-average development level, and possibly way up there among the most advanced....[W]e may be one of the most advanced civilizations anywhere."

The book's final chapter is an effort to prove that the probability of life on at least one other planet, though not absolutely certain, is so close to 1 that we can assume certainty. This argument requires a long chain of probability questions that are not, as mathematicians like to say, well-formed. It seems likely there are billions of planets in every galaxy, but astronomers do not yet know how probable it is that a planet meets all the conditions necessary for life. A planet must be neither too hot nor too cold (Aczel calls this the Goldilocks criterion). Does it have water? A suitable atmosphere? How likely is it that carbon is present? Because we do not know the probability that a planet will have all these features, it is impossible, in my opinion, to make a reliable estimate of the chances that conditions on other planets will be conducive to life.

Even if all necessary environmental conditions are present, no one has the slightest notion of how to estimate the probability that carbon molecules, jiggling in a primordial soup, will form self-replicating molecules. It may turn out, as more is learned, that under ideal conditions, it is almost certain that DNA molecules will appear. Or it may turn out, as antipluralists insist, that the chance of this happening is so close to zero that it is extremely unlikely that life has arisen anywhere else in our galaxy, perhaps not even in other galaxies.

The view that we are alone in the universe will, of course, be dealt a near-fatal blow if our space probes find some form of life on Saturn's moon Titan, so rich in atmosphere, or in the seas below the surface ice of Jupiter's moon Europa. But even if there are low forms of life on Titan or Europa, they will tell us nothing about the probability that given enough time, these life forms will evolve into creatures as intelligent as ourselves. "A sad spectacle!" wrote Thomas Carlyle as he considered the possibility of extraterrestrial life on other planets. "If they be inhabited, what a scope for pain and folly; and if they be not inhabited, what a waste of space!"

I for one would be delighted, albeit a bit terrified, if tomorrow we received a radio signal from outer space that repeated the counting numbers or the prime numbers. As several scientists have warned: "If we hear such a message, don't answer!" I am equally undismayed by the prospect that Earth is the only planet with life (although we could never be sure of this) and that we are destined to colonize our galaxy. It is said that William Randolph Hearst once cabled a famous British astronomer to request 1000 words on the question "Is there intelligent life elsewhere in the universe?" Hearst got back an answer in which "Don't know" was repeated five hundred times.

I, too, have often felt the emotions of that anonymous cowboy who wrote "Home on the Range":

How often at night,
when the heavens are bright
with the light from the glittering stars,
have I stood there amazed
and asked as I gazed,
does their glory exceed that of ours?

Chapter 41
Fuzzy New New Math

*Multicultural and Gender Equity in the Mathematics Class-
room: The Gift of Diversity* (1997 Yearbook); edited by Janet
Trentacosta and Margaret J. Kenney. National Council of
Teachers of Mathematics, 248 pp., $22.00

Focus on Algebra: An Integrated Approach by Randall I.
Charles, Alba Gonzalez Thompson, et al. Addison-Wesley,
843 pp., $56.00

Life by the Numbers: Math As You've Never Seen It Before
narrated by Danny Glover. Seven boxed videotapes produced by WQED, Pittsburg, $129.00

Surveys have shown for many decades that the mathematical skills
of American high school students lag far behind those of their coun-
terparts in Japan, Korea, Singapore, and many European countries.
In the United States whites do better than blacks, Hispanics, and
Native Americans. Males outscore females. Students from high so-
cioeconomic backgrounds do better than those from lower strata.

These are troubling statistics because, in an advanced technologi-
cal society such as ours, a firm grasp of basic mathematics is increasingly essential for better-paying jobs. Something clearly is wrong with
how math is being taught in precollege grades, but what?

In the late 1960s the National Council of Teachers of Mathematics
(NCTM) began to promote a reform movement called the New Math.
In an effort to give students insight into why arithmetic works, it
placed a heavy emphasis on set theory, congruence arithmetic, and
the use of number bases other than ten. Children were forbidden to
call, say, 7 a "number." It was a "numeral" that symbolized a number.

This review appeared in *The New York Review of Books* (September 24, 1998).

The result was enormous confusion on the part of pupils, teachers, and parents. The New Math fad faded after strong attacks by the physicist Richard Feynman and others. The final blow was administered by the mathematician Morris Kline's 1973 best seller *Why Johnny Can't Add: The Failure of the New Math*.

Recently, the NCTM, having learned little from its New Math fiasco, has once more been backing another reform movement that goes by such names as the new new math, whole math, fuzzy math, standards math, and rain forest math. Like the old New Math, it is creating a ferment among teachers and parents, especially in California, where it first caught on. It is estimated that about half of all precollege mathematics in the United States is now being taught by teachers trained in fuzzy math. The new fad is heavily influenced by multiculturalism, environmentalism, and feminism. These trends get much attention in the twenty-eight papers contributed to the NCTM's 1997 yearbook, *Multicultural and Gender Equity in the Mathematics Classroom: The Gift of Diversity*.

It is hard to fault most of this book's advice, even though most of the teachers who wrote its chapters express themselves in mind-numbing jargon. "Multiculturalism" and "equity" are the book's most-used buzzwords. The word "equity," which simply means treating all ethnic groups equally, and not favoring one gender over another, must appear in the book a thousand times. A typical sentence opens Chapter Eleven: "Feminist pedagogy can be an important part of building a gender-equitable multicultural classroom environment." Over and over again teachers are reminded that if they suspect blacks and females are less capable of understanding math than Caucasian males, their behavior is sure subtly to reinforce such beliefs among the students themselves, or what one teacher calls, in the prescribed jargon, a student's "internalized self-image."

"Ethnomathematics" is another popular word. It refers to math as practiced by cultures other than Western, especially among primitive African tribes. A book much admired by fuzzy-math teachers is Marcia Ascher's *Ethno-mathematics: A Multicultural View of Mathematical Ideas* (1991).[1] "Critical-mathematical literacy" is an even longer jawbreaker. It appears in the NCTM yearbook as a term for the ability to interpret statistics correctly.

Knowing how pre-industrial cultures, both ancient and modern, handled mathematical concepts may be of historical interest, but one must keep in mind that mathematics, like science, is a cumulative process that advances steadily by uncovering truths that are everywhere the same. Native tribes may symbolize numbers by using different base systems, but the numbers behind the symbols are identical. Two elephants plus two elephants makes four elephants in every African tribe, and the arithmetic of these cultures is a miniscule portion of the vast jungle of modern mathematics. A Chinese mathematician is no more concerned with ancient Chinese mathematics, remarkable though it was, than a Western physicist is concerned with the physics of Aristotle.

Fuzzy-math teachers are urged by contributors to the yearbook to cut down on lecturing to passive listeners. No longer are they to play the role of "sage on stage." They are the "guide on the side." Classes are divided into small groups of students who cooperate in finding solutions to "open-ended" problems by trial and error. This is called "interactive learning." The use of calculators is encouraged, along with such visual aids as counters, geometrical models, geoboards, wax paper (for folding conic section curves), tiles of different colors and shapes, and other devices. Getting a correct answer is considered less important than shrewd guesses based on insights, hence the term "fuzzy math." Formal proofs are downgraded.

No one can deny the usefulness of visual aids. Teachers have known for centuries that the best way to teach arithmetic to small children is by letting them "interact" with counters. Each counter models anything that retains its identity—an apple, cow, person, star. What's the sum of 5 and 2? A girl who knows how to count moves into a pile five counters, then two more, and counts the heap as seven. Suppose she first moves two, then five. Does it make a difference? Similar procedures teach subtraction, multiplication, and division.

After a few days of counter playing it has been traditional for children to memorize the addition table to at least 9. Later they learn the multiplication table to at least 10. "Hands-on" learning first, then rote learning. Unfortunately, some far-out enthusiasts of new new math reject anything resembling what they call "drill and kill" memorizing. The results, of course, are adults who can't multiply 12 by 12 without reaching for a calculator.

Aside from its jargon, another objectionable feature of the yearbook is that its contributors seem wholly unaware that the best way to keep students awake is to introduce recreational material that they perceive as fun. Such material includes games, puzzles, magic tricks, fallacies, and paradoxes. For example, determining whether the first or second player can always win at tic-tac-toe, or whether the game is a draw if each player makes the best moves, is an excellent way to introduce symmetry, combinatorics, graph theory, and game theory. Because all children know the game, it ties strongly into their experience.

For what the yearbook likes to call a "cognitively challenging" task, give each child a sheet with a checkerboard on it. Have each of them cut off two opposite corner squares. Can the remaining sixty-two squares be covered by thirty-one dominoes? After a group finds it impossible, see how long it takes for someone to come up with the beautiful parity (odd–even) proof of impossibility.

If new new math teachers are aware of such elegant puzzles, and there are thousands, there is no hint of it in the yearbook. This is hard to understand in view of such best-selling textbooks as Harold Jacobs's *Mathematics: A Human Endeavor* (1970; third edition, 1994), which has a great deal of recreational material; *Mathematics: Problem Solving Through Recreational Mathematics*, a textbook by Bonnie Averbach and Orin Chein (1980); and scores of recent books on entertaining math by eminent mathematicians.

I seldom agree with the conservative political views of Lynne Cheney, but when she criticized extreme aspects of the new new math on the Op-Ed page of *The New York Times* on August 11, 1997,[2] I found myself cheering. As Cheney points out, at the heart of fuzzy-math teaching is the practice of dividing students into small groups, then letting them discover answers to problems without being taught how to find them. For example, teachers traditionally introduced the Pythagorean theorem by drawing a right triangle on the blackboard, adding squares on its sides, and then explaining, perhaps even proving, that the area of the largest square exactly equals the combined areas of the two smaller squares.

According to fuzzy math, this is a terrible way to teach the theorem. Students must be allowed to discover it for themselves. As Cheney describes it, they cut from graph paper squares with sides ranging from two to fifteen units. (Such pieces are known as "manipulatives.") Then they play the following "game." Using the edges of the squares, they form triangles of various shapes. The "winner" is the first to discover that if the area of one square exactly equals the combined areas of the other two squares, the triangle must have a right angle with the largest square on its hypotenuse. For example, a triangle of sides 3, 4, 5. Students who never discover the theorem are said to have "lost" the game. In this manner, with no help from teacher, the children are supposed to discover that with right triangles $a^2 + b^2 = c^2$. "Constructivism" is the term for this kind of learning. It may take a group several days to "construct" the Pythagorean theorem. Even worse, the paper game may bore a group of students more than hearing a good teacher explain the theorem on the blackboard.

One of the harshest critics of fuzzy math is the writer John Leo, whose article on the subject, "That So-Called Pythagoras," was published last year in *US News and World Report* (May 26, 1997). (His title springs from a reference he found in a book on ethno-mathematics to "the so-called Pythagorean theorem.") Leo tells of Marianne Jennings, a professor at Arizona State University, whose daughter was getting an A in algebra but had no notion of how to solve an equation. After obtaining a copy of her daughter's textbook, Jennings soon understood why. Here is how Leo describes this book:

> It includes Maya Angelou's poetry, pictures of President Clinton and Mali wood carvings, lectures on what environmental sinners we all are and photos of students with names such as Taktuk and Esteban "who offer my daughter thoughts on life." It also contains praise for the wife of Pythagoras, father of the Pythagorean theorem, and asks students such mathematical brain teasers as "What role should zoos play in our society?" However, equations don't show up until Page 165, and the first solution of a linear equation, which comes on Page 218, is reached by guessing and checking.

Romesh Ratnesar's article "This is Math?" (*Time*, August 25, 1997) also criticizes the new new math. It describes fifth-graders who were asked how many handshakes would occur if everyone in the class shook hands with everyone else. At the end of an hour, no group had the answer. Unfazed, the teacher said they would be trying again after lunch. Professor Jennings makes another appearance. She told Ratnesar that she became angry and worried when she saw her daughter use her calculator to determine 10 percent of 470.

Curious about her daughter's textbook, which is now widely used, I finally obtained a copy by paying a bookstore $59.12. Titled *Focus on Algebra: An Integrated Approach*, this huge text contains 843 pages and weighs close to four pounds. (In Japan, the average math textbook is two hundred pages.) It is impossible to imagine a sharper contrast with an algebra textbook of fifty years ago.

"Integrated" in the subtitle has two meanings:

(1) Instead of being limited to algebra, the book ranges all over the math scene with material on geometry, combinatorics, probability, statistics, number theory, functions, matrices, and scatter graphs, and of course the constant use of calculators and graphers. Fifty years ago high school math was given in two classes, one on algebra, one on geometry. Today's classes are "integrated" mixtures.

(2) The book is carefully integrated with respect to gender and to ethnicity, with photographs of girls and women equal in number to photographs of boys and men. Faces of blacks and whites are similarly equal, though I noticed few faces of Asians.

On the positive side is the book's lavish use of color. Only a few pages lack full-color photos and drawings, all with eye-catching layouts. When it comes to actual mathematics the text is for the most part clear and accurate, with a strong emphasis on understanding why procedures work, and on inducements to think creatively. "After all," the text says on its first page, "what good is it to solve an equation if it is the wrong equation?" The trouble is that the book's mathematical content is often hard to find in the midst of material that has no clear connection to mathematics.

Not having taught mathematics myself, I have no opinion about the value of students working in small groups as opposed to sitting and listening to a teacher talk. Nor have I found research studies that make a decisive case in favor of either method. Clearly a great deal depends on the qualities of particular teachers, and these would be hard to appraise in any survey. The authors justify the group approach by saying it anticipates the workplaces in which students will find themselves as adults. John Donne's remark about how no man is an island is quoted. The book's first "exercise" is a question: "In general, do you prefer to work alone or in groups?"

An emphasis on ethnic and gender equity is, of course, admirable, though in this textbook it seems overdone. For example, twelve faces of boys and girls of mixed ethnicity reappear in pairs throughout the pages. Each has something to say. "Taktuk thinks..." is followed with "Esteban thinks...," "Kirti thinks..." is followed by what "Keisha thinks...," and so on. These pairings become mechanical and predictable.

The book jumps all over the place, with transitions as abrupt as the dream episodes of *Alice in Wonderland*. I think most students would find this confusing. Eight full pages are devoted to statements by adult professionals, with their photographs. Each statement opens with a sentence about whether they liked or disliked math in high school, followed by generally banal remarks. For example, Diana Garcia-Prichard, a chemist, writes: "I liked math in high school because all the problems had answers. Math is part of literacy and the framework of science. For instance, film speed depends on chemical reactions. I use math to model problems and design experiments. I like getting results that I can publish and share." Presumably such statements are intended to convince students that math will be useful later on in life.

Many of the book's exercises are trivial. For example, on page 20 students are asked to play forward and backward a VCR tape of a skier, then answer the question: "How will this affect the way the skier appears to move?" On page 11: "A circle graph represents 180 kittens. What does 1/4 of the circle represent?" (Answer, to be found in the back of the book: 45.) A chapter on "the language of algebra" opens with a page on the origin of such phrases as "the lion's share," "the boondocks," and "not worth his salt." It is not clear what this has to do with algebra.

Many pictures have only a slim relation to the text. Magritte's painting of a green apple floating in front of a man's face accompanies some problems about apples. Van Gogh's self-portrait is alongside a problem about the heights and widths of canvases. A picture of the Beatles accompanies a problem about taxes only because of the Beatles' song "Taxman." My favorite irrelevant picture shows Maya Angelou talking to President Clinton. Beside it is the following extract from one of her prose poems:

> Lift up your eyes upon
> This day breaking for you
> Give birth again
> To the dream.
> Women, Children, Men,
> *Take it into* the palm of your hands.
> *Mold it into* the shape of your most
> Private need. *Sculpt it into*
> The image of your most public self.

Why is this quoted? Because the "parallel" phrases shown underlined are similar to parallel lines in geometry! Is this intended to "integrate" geometry and poetry?

The book is much concerned with how the environment is being polluted. Protecting the environment is obviously a good cause, but here its connection with learning math is often oblique, if not arbitrary. A chapter on functions opens with a page headed "Unstable Domain." Its first question is "What other kinds of pollution besides air pollution might threaten our planet?" Page 350 has a picture of crude oil being poured over a model of the earth. It accompanies a set of questions relating to the way improper disposals of oil are contaminating ground water.

A page headlined "Hot Stuff" shows three kinds of peppers to illustrate how they are used in cooking. Two of the "exercises" are: "The chili cook-off raises money for charity. Describe some ways the organizers could raise money in the cook-off," and "How would you set up a hotness scale for peppers?" This page introduces a chapter on how to solve linear equations.

Another section on equations opens with pictures of zoo animals. It discusses what can be done to prevent species from becoming extinct. The first question is "What role should zoos play in today's society?" The book's index, under the entry "Animal study and care," lists thirty-two page references.

A section on mathematical inequalities is preceded by a page on how Mary Rodas became vice-president of a toy company, and how Linda Johnson Rice found a creative way to market Eboné cosmetics for black women. Under a photo of a smiling Mary, the first questions are: "Would you like to own your own business someday? Why or why not?"

On page 67, a picture of Toni Morrison is used to illustrate a problem about how many ways four objects can be placed in a row. The text then introduces four students who each read an excerpt from something Morrison has written. In how many different orders, the text asks, can the four excerpts be read? A man from Mysore, India, who creates shadow pictures on the wall with his fingers is featured on page 421. What this has to with the following section on solving systems of inequalities is not evident. A photo of Alice Walker on page 469 illustrates the question: "Is the time it takes to read an Alice Walker novel always a function of the number of pages?" This and other such references give the impression that well-known writers are being dragged into the text.

The most outrageous page—it opens a section on linear functions—concerns the Dogon culture of West Africa. Students are told that this primitive tribe, without the aid of telescopes, discovered that Jupiter has satellites, that Saturn has rings, and that an invisible star of great density orbits Sirius once every fifty years. Presumably the Dogon had supernormal powers. It has long been known, however, that the Dogon made no such discoveries. They merely learned these astronomical facts from missionaries and other Western visitors.[3]

Like the authors of the NCTM yearbook, those who fashioned this huge textbook seem wholly uninterested in recreational material. The book's only magic trick (page 246) is a stale, utterly trivial way to guess a number. Although strongly favoring the use of calculators, the authors don't seem aware that the hundreds of amazing number tricks that can be done with them provide excellent exercises. A child can learn a lot of significant number theory in discovering why they work. None is in the book.

An old brain teaser involves a glass of wine and a glass of water. A drop is taken from the wine and added to the water. The water is stirred, then a drop of the mixture goes back to the wine. Is there now more water in the wine than wine in the water, or vice versa? The surprising answer is that the two amounts are precisely equal.

Students will be fascinated by the way this principle can be modeled with a deck of cards. Divide the deck in half, one half consisting of all the red cards, the other half consisting of all the black cards. Take as many cards as you like from the red (wine) half and insert them anywhere among the blacks (water). Shuffle the black half. From it remove from anywhere the same number of cards you took from the reds and put them back among the reds. You'll find the number of blacks among the reds is exactly the same as the number of reds among the blacks. Students will enjoy proving that this is always the case. But will it work if the two starting portions of the deck are unequal? (Yes. It doesn't matter if the two glasses in the brain teaser are not the same size; nor does it matter how many cards are in the black and red piles.)

This secondary math textbook has an index that is not very helpful. What value are more than 180 page references for the entry "Science"? What use is a similar quantity of page numbers for the entry "Industry"?

WQED's boxed set of seven video-tapes, *Life by the Numbers*, was funded mainly by the National Science Foundation and the Alfred P. Sloan Foundation. The photography is superb. There are scenes of men and women mathematicians seated at computer consoles, driving cars, or walking down a street or through the woods. There are many close-ups of their faces, dazzling glimpses of mountains and skyscrapers, baseball games, martial arts contests, blossoming flowers, wild animals, and everything imaginable that has little to do with math. The tapes rate high on special effects, low on mathematical content.

The seventh tape covers a typical new new math class. To discover that the longer a pendulum, the slower its swing, students tie weights to the ends of string and swing the weights back and forth while other students keep charts of string lengths and pendulum periods. After several days they learn that the period of a pendulum is a function of its length. This discovery enables them to calculate whether the victim in Poe's horror story "The Pit and the Pendulum" has enough time to escape from the huge pendulum which threatens to cut him in half as it swings lower and lower over his reclining body. It is assumed that because students have fun swinging weights they will remember the function better than if a teacher takes a few minutes to demonstrate it by swinging a weight and slowly lengthening the string.

One of the most telling attacks on new new math is Bernadette Kelly's article "Déjà Vu? The New 'New Math,'" in the professional journal *Effective School Practices* (Spring 1994). Kelly summarizes four case studies by four supporters of fuzzy math in which they report on four fifth-grade teachers.[4] Two of the teachers, called Sandra and Valerie, are enthusiastic users of new new math techniques. The other two teachers use traditional methods.

Sandra was very good at getting students to cooperate in groups. However, in one exercise she told students that one could obtain the perimeter of a rectangular field by multiplying its length by its width! In another project she calculated the volume of a sandbox by multiplying together its length and width in yards, then multiplying the product by the box's height in feet!

In an interview Sandra said that while working on the sandbox problem her pupils asked what a cubic foot was. "You know, the thing is that I couldn't really answer that question. Then I thought and thought, then I remembered how to measure a cube." Neither Sandra nor her students were ever aware of her two huge mistakes. In spite of these errors, the author of the article about her said she was an "exemplary teacher." Sandra is praised for getting her students to enjoy their cooperative efforts to solve problems "in the context of real world situations." Finding a correct answer was less important than having fun in working on the problem.

Valerie made an equally astonishing blunder. The task was to determine the average number of times her thirty students had eaten ice cream over a period of eight days. This was "solved," by dividing 30 by 8, to get 3.75, which Valerie rounded up to 4!

As with Sandra, neither Valerie nor her students ever became aware that they obtained a totally wrong answer. Nevertheless, the author of the paper about her forgives her mistake on the grounds that she had succeeded so well in getting her students to work on a problem in the context of their experience. Moreover, the work had impressed on the students the "usefulness and relevance of averages." No matter that they completely failed to find an average.

As for Jim and Karen, the two teachers who used more traditional methods, the authors of the case studies are unimpressed by their students having scored high on tests. Both are castigated for failing to appreciate the methods of the new new math. What is deplorable, as Bernadette Kelly's article points out, is not so much that the case studies revealed the incompetence of two teachers, who come through as ignoramuses, as the authors' praise of Sandra and Valerie for finding ways to get their pupils working joyfully on problems. Little wonder that new new math is called fuzzy. Insights are deemed significant even when they are wrong.

The mathematician Sherman Stein, in his 1996 book *Strength in Numbers*, devotes a chapter to a history of math reform movements. His hopes for the new new math are dim. "I am disturbed," he writes,

> that the authors of the [new new math books] do not cite any pilot project or any school district as a model to show that their goals can be achieved in the real world. That means that they are proposing to change the way an entire generation learns mathematics without checking the feasibility of their recommendations. A manufacturer introduces a new soap with more care, first testing its reception in a few stores or towns before committing to mass production.

But evaluating the efficacy of fuzzy math will not be easy. Too many variables are involved, including the skill of teachers and the educational background of parents, to mention only two. A glaring example of how research can be biased is provided by a recent testing of precollege math students around the world by the Third International Mathematics and Science Study. Results announced last. February revealed that American students did better than students in just two other countries, Cyprus and South Africa. A cartoon in *The New York Times* (March 8) showed a car's bumper sticker that said "My kid's math scores beat kids in Cyprus and South Africa." Inside the car the son is giving a thumbs-up sign.

These statistics are worthless. In many cases the students in a foreign country were much older than students here at the same grade level. More significantly, in most foreign nations students in early grades who show no aptitude for math are sent off to trade schools or to jobs, if they can find them. In the US such students are required to continue attending high school. Obviously our high school students will do less well on math tests than students in countries where poor students are quickly moved out of the system.

Although we lack clear, systematic evidence that methods of fuzzy math are inferior to older methods, education officials in California, the nation's largest customer for math textbooks, have suddenly turned against the new new math. The change in state policy was mainly in response to the outrage of parents who complained that their children were unable to do the simplest arithmetical calculations. Their outrage was backed by many top mathematicians and scientists. Michael McKeown, for example, a distinguished molecular biologist at the Salk Institute, heads a parental group called Mathematically Correct. "We're not opposed to teaching concepts," he told *Newsweek* ("Subtracting the New Math," December 15, 1997). "I am opposed to failing to give a kid tools to solve a problem."

In a vote of ten to zero (one person abstained) the eleven members of California's Board of Education recommended this spring a broad return to basics in math teaching. The decision is sure to have an effect in other states. The board said students should learn the multiplication table by the end of the third grade, and that fourth-graders should know how to do long division without consulting a calculator. It banned the use of calculators on state tests. Teachers were urged not to introduce calculators before grade six.

Defenders of fuzzy math are, of course, dismayed. They branded the board's decisions a product of nostalgia, and a contribution to our country's dumbing down. The National Science Foundation, which has given more than $50 million to California districts for research on new math teaching, is furious. It has threatened to withdraw further funding to any California district that adopts the board's recommendations.

The conflict is bitter and far from over. It may be many years before it becomes clear how to sift out from the new new math what is valuable while retaining worthy aspects of older teaching methods.[5] My own opinion is that the most important question concerning the teaching of math is not how big and colorful textbooks are, how many visual aids are used, how the classroom is physically arranged, or even what methods are used in it. The greatest threat to good math teaching is surely the low pay that keeps so many excellent teachers and potential teachers out of our schools. What matters more than anything else is having trained teachers who understand and love mathematics, and are capable of communicating its mystery and beauty tp their pupils.

Notes

1. Textbooks emphasizing multiculturalism are proliferating rapidly. Here are a few: *Africa Counts: Number and Pattern in African Culture*, by Claudia Zaslavsky (Lawrence Hill, 1997); *Multiculturalism in Mathematics, Science and Technology*, by Miriam Barrios-Chacon and others (Addison-Wesley 1993); *Multicultural Mathematics: Teaching Mathematics from a Global Perspective*, by David Nelson, George Gheverghese Joseph, and Julian Williams (Oxford University Press, 1993); *Teaching with a Multicultural Perspective: A Practical Guide*, by Leonard Davidman and Patricia T. Davidman (Perseus, 1996). Striking multicultural math posters are available from teaching supply houses.

2. See also the letters in *The New York Times* of August 17, 1997, and an earlier article by Cheney in the *Weekly Standard* (August 4, 1997).

3. On the myth of Dogon astronomy, see Carl Sagan, *Broca's Brain* (Random House, 1979), pp. 63–64 and Chapter Six; Ian Ridpath, "Investigating the Sirius 'Mystery,'" in *The Skeptical Inquirer*, Vol. 3 (Fall 1978), pp. 56–62; and Terence Hines, *Pseudoscience and the Paranormal* (Prometheus Books, 1988), pp. 216–219.

4. R.T. Putnam, R.M. Heaton, R.S. Prewat, and J. Remillard, "Teaching Mathematics for Understanding," in *Elementary School Journal*, Vol. 93 (1992), pp. 213–228.

5. That the new new math has positive aspects goes without saying. It is important that students understand the basic concepts of math and not just memorize procedures that work; and to give students such conceptual understanding teachers themselves must have such understanding. This is the theme of a recent monograph, *Middle Grade Teachers' Mathematical Knowledge and Its Relationship*

to Instruction, by Judith Sowder, Randolph Philipp, Barbara Armstrong, and Bonnie Schappelle (State University of New York, 1998).

The monograph reports on a two-year investigation of five teachers, with a primary emphasis on how they taught fractions. Why, for example, in dividing one fraction by another do you flip upside down the divisor fraction, then multiply the numerators and denominators? Should teachers be content with letting students accept this as a trick that works like magic, or try to answer a student who asks, "Why is this division?" The monograph defends the admirable aspects of math reform without going to fuzzy-math extremes.

An anonymous piece of humor circulating on the internet purports to give capsule summaries of how a simple problem is handled at four stages in the recent history of mathematics teaching in the United States.

The problem: A logger sells a truckload of lumber for $100. His production cost is 4/5 of the price. What is his profit?

1960, the old math wanes...

A logger sells a truckload of lumber for $100. His cost of production is four-fifths of the price, or $80. What is his profit?

1970, the new math...

A logger exhanges a set "L" of lumber for a set "M" of money. The cardinality of set "M" is 100, and each element is worth $1. Make 100 dots representing the elements of the set "M". The set "C" of the costs of production contains 20 fewer points than set "M". Represent the set "C" as a subset of "M", and answer the following question: What is the cardinality of the set "P" of profits?

1980, leveling the playing field...

A logger sells a truckload of lumber for $100. His cost of production is $80, and his profit is $20. Your assignment: underline the number 20.

1990, The New New Math...

By cutting down beautiful forest trees, a logger makes $20. What do you think of this way of making a living? Topic for class participation: How did the forest birds and squirrels feel?

Postscript

The following two letters appeared in *The New York Review of Books* (December 3, 1998), with my reply:

★   ★   ★

There are many aspects of Martin Gardner's critique of the current reform movement in mathematics education that invite rebuttal. In this communication I would like to take up his treatment of the Pythagorean theorem.

Gardner objects to having students discover this important mathematical principle through the use of "manipulatives." He describes a game in which students "cut from graph paper squares with sides ranging from two to fifteen units.... Using the edges of the squares, they form triangles of various shapes." The "winner," he says, "is the first to discover that if the area of one square exactly equals the combined areas of the other two squares, the triangle must have a right angle with the largest square on its hypotenuse.... 'Constructivism' is the term for this kind of learning. It may take a group several days to 'construct' the Pythagorean theorem. Even worse, the paper game may bore a group of students more than hearing a good teacher explain the theorem on the blackboard."

I teach a college-level geometry course for future elementary teachers in which we spend several days studying the Pythagorean theorem and analyzing a few of the four hundred known proofs of this important proposition. At the end of this experience, I ask students to write in their journals about any new insights they have obtained from this reexamination of the Pythagorean theorem.

More often than not, students write something along these lines. "I always knew that the Pythagorean theorem was expressed by the formula $a^2 + b^2 = c^2$, but I was surprised to learn that the theorem actually describes a relationship among areas of squares."

Why is it that these college students find the connection between the Pythagorean theorem and the areas of squares such a revelation? From my experience most high school geometry textbooks prove the theorem using properties of similar right triangles. While this is one of the shorter proofs, it obscures the fact that the terms in the formula $a^2 + b^2 = c^2$ represent areas of actual squares. Too often students simply memorize the formula without any real understanding of what a, b, and c stand for.

I wish our students would come to the university having learned the Pythagorean theorem through the method that Gardner ridicules. The activity he describes gives students the opportunity to work together and look for patterns. In the process they may discover more than just the Pythagorean theorem. They should also observe that in obtuse triangles the square of the longest side is greater than, and in acute triangles it is less than, the sum of the squares of the other two sides. They should also observe or rediscover that the sum of the lengths of any two sides must be greater than the length of the third side.

These discoveries, however, cannot be left to chance, but are more likely to take place under the guidance of a teacher. Contrary to Gardner's assertion, cooperative learning does not imply that children receive "no help" from the teacher, nor that there are "winners" who discover what they are supposed to and "losers" who don't. A skilled teacher guides the entire class in a discussion of what they have observed and in formulating the appropriate generalizations. Gardner is correct to observe that some students will become bored if the "exploration" phase is prolonged just as many students have been bored by brilliant lectures delivered by "good teachers." The art of teaching involves determining an appropriate mixture of teacher-led and student-directed investigation and in determining when to intervene in cooperative group discussions.

Contrary to its most vocal critics, the mathematics education reformers do not advocate that teachers abdicate their roles as leaders in the classroom, nor that we abandon precision and rigor. It is ironic that Gardner accuses the reform movement of promoting "fuzzy" math, when traditional methods of teaching seem to result in so much fuzzy understanding among our students.

Timothy V. Craine
Associate Professor
Department of Mathematical Sciences
Central Connecticut State University
New Britain, Connecticut

* * *

I am the author of the case of "Sandra," one of the cases of fifth-grade teachers Martin Gardner discusses and quotes from. Gardner draws on his knowledge of Sandra and the other three teachers from two sources. One source is Bernadette Kelly's article (*Effective School Practices*, Spring 1994) and the other is a summary article based on the four cases (*Elementary School Journal*, Volume 93, 1992). The original four cases of Sandra, Valerie, Jim, and Karen, which were written and published as four individual articles (*Elementary School Journal*, Volume 93, 1992), are not referenced by Gardner yet he criticizes the authors and texts of these cases. For example, I am criticized by Gardner for saying Sandra was an "exemplary teacher." In the text of my case, I accurately reported from my sources that "Sandra likes to teach mathematics and is identified by both district personnel and a university-based staff developer as someone who teaches in ways consistent with the Framework. She is, from their perspective, an exemplary teacher within her school district" (p. 155). The label, "exemplary," was one given to Sandra by sources I interviewed, not one I attributed to her. Gardner's other criticisms of my work are based on statements from my case pulled out of context and similar to those used by Kelly (*Effective School Practices*, Spring 1994).

As a qualitative educational researcher, it is my responsibility to report what I see and to try to understand what I see from the participant's point of view. Sandra's practice seemed outrageous and perplexing to me as an observer. However, in writing the case, my aim was to try to explain what I observed. Gardner states, "What is deplorable, as Bernadette Kelly's article points out, is not so much that the case studies revealed the incompetence of two teachers,... as the authors' praise of Sandra and Valerie for finding ways to get their pupils working joyfully on problems." What Gardner and Kelly interpret as praise is my effort as a researcher not to bash a teacher and to try to understand what I observed from Sandra's perspective. In the text of my case, I state, "My observations in Sandra's classroom illustrate what can happen when a teacher who has the best of intentions tries to teach a concept that he or she does not understand. Although subject matter knowledge alone is insufficient for teachers to teach for understanding, it is nonetheless crucial. Sandra's case dramatizes the role of subject matter knowledge—its need becomes evident through the consequences of its absence" (p. 161). The title of my case, "Who is Minding the Mathematics Content? A Case Study of a Fifth Grade

Teacher," also represents my concern over attention to subject matter in elementary mathematics instruction. Had Gardner read the original text of my case of Sandra, I think he would have seen that he and I agree on the need for elementary teachers to understand the mathematics they are aiming to teach.

Ruth M. Heaton, Ph.D.
Assistant Professor
University of Nebraska–Lincoln
Lincoln, Nebraska

★　★　★

I replied as follows:

I am even more astounded than Mr. Craine to learn that college students in a math class would not know that the Pythagorean theorem is about geometrical squares. Surely this testifies to the miserable quality of pre-college math teaching.

The question is how best to teach this theorem to children in lower grades. Class time is precious. I cannot believe that a group of youngsters, struggling for days to invent the Pythagorean theorem by manipulating paper cutouts, will remember the theorem any better than hearing a good teacher explain it on a blackboard.

The University of Indiana's mathematician/logician Raymond Smullyan once taught high school geometry. He would chalk a right triangle on the blackboard and its three squares. Assuming the squares are made from sheets of gold, he would then ask his students which would they rather have: the large gold square or the two smaller squares. His students would be flabbergasted when told it made no difference because regardless of the right triangle's shape the big square's area always equals the total area of the two smaller ones.

I think this demonstration of the theorem would become more firmly fixed in a child's mind than spending days in a group trying (perhaps unsuccessfully) to discover the theorem by themselves. Moreover, it allows the teacher time to explain how the theorem applies to semicircles and other similar shapes erected on the triangle's sides, and other fascinating related facts. As a homework assignment, students could be asked to search for one of the hundreds of proofs mentioned by Craine. Unfortunately, fuzzy math educators are not much interested in formal proofs.

It is true that my comments on Ruth Heaton's article about Sandra were based entirely on Bernadette Kelly's paper. I am pleased that Ms. Heaton is given this opportunity to clarify her opinion of Sandra's failure to understand the math she was supposed to teach.

Printed in the United States
by Baker & Taylor Publisher Services